Undersea Treasures

Prepared by the Special Publications Division
National Geographic Society
Washington, D. C.

UNDERSEA TREASURES

Contributing Authors
SUE SWEENEY ABBOTT, ROBERT E. CAHILL, WILLIAM
  GRAVES, MENDEL PETERSON, ROBERT STÉNUIT, PETER
  THROCKMORTON

Published by
THE NATIONAL GEOGRAPHIC SOCIETY
MELVIN M. PAYNE, *President*
MELVILLE BELL GROSVENOR, *Editor-in-Chief*
GILBERT M. GROSVENOR, *Editor*
BART MCDOWELL, *Consulting Editor*

Prepared by
THE SPECIAL PUBLICATIONS DIVISION
ROBERT L. BREEDEN, *Editor*
DONALD J. CRUMP, *Associate Editor*
PHILIP B. SILCOTT, *Senior Editor*
RONALD M. FISHER, *Managing Editor*
MENDEL PETERSON, *Consultant, Director of Underwater
  Exploration, retired, Smithsonian Institution*
JAN NAGEL CLARKSON, STRATFORD C. JONES, TOM
  MELHAM, TEE LOFTIN SNELL, *Assistants to
  the Editor*
TEE LOFTIN SNELL, *Research*

Illustrations
GEORGE PETERSON, *Picture Editor*
SUSAN C. BURNS, JAN NAGEL CLARKSON, LOUIS DE LA
  HABA, BARBARA GRAZZINI, STRATFORD C. JONES,
  TOM MELHAM, WENDY VAN DUYNE,
  Picture Legends
MARJORIE W. CLINE, LOUISA MAGZANIAN, *Picture
  Research*

Design and Art Direction
JOSEPH A. TANEY, *Staff Art Director*
JOSEPHINE B. BOLT, *Associate Art Director*
URSULA PERRIN, *Staff Designer*
JOHN D. GARST, JR., MARGARET A. DEANE, NANCY
  SCHWEICKART, MILDA R. STONE, ROBERT J. VILSECK,
  Map Research and Production

Production and Printing
ROBERT W. MESSER, *Production Manager*
GEORGE V. WHITE, *Assistant Production Manager*
RAJA D. MURSHED, NANCY W. GLASER, *Production
  Assistants*
JOHN R. METCALFE, *Engraving and Printing*
MARY G. BURNS, JANE H. BUXTON, MARTA ISABEL
  COONS, SUZANNE J. JACOBSON, PENELOPE A.
  LOEFFLER, SANDRA LEE MATTHEWS, JOAN PERRY,
  MARILYN L. WILBUR, *Staff Assistants*
VIRGINIA S. THOMPSON, JOLENE MCCOY, *Index*

**Overleaf: Twenty-five feet down in the North Sea,
treasure diver Robert Sténuit examines brass navigational dividers extracted from concretions on the
ocean floor. A thick bed of kelp hides wreckage of the
Dutch ship Lastdrager, sunk in 1653. Page 1: A six-
pound gold disk, recovered from a Spanish treasure
galleon, bears royal tax stamps. Raised from the same
ship, off Key West, Florida, the gold chain measures
8½ feet. Hard cover: Trailing black coral, a diver
surfaces from 200 feet down in the Bahamas.**

DAVID DOUBILET (ABOVE); OVERLEAF AND PAGE 1: NATIONAL
GEOGRAPHIC PHOTOGRAPHER BATES LITTLEHALES; HARD COVER
ADAPTED FROM A PHOTOGRAPH BY BATES LITTLEHALES

In the Caribbean, an amateur diver bobs near a boat holding

Foreword

IN MY YEARS OF DIVING, sailing, and exploring the oceans, I've found *one piece* of
sunken gold, a Spanish doubloon lost at sea
some 250 years ago. That find—made in
1953 off Florida's Tavernier Key with Art
McKee and Mendel Peterson—infected me
with an instant case of gold fever. But when
weeks of search failed to turn up the slight-
est trace of more gold, my fever quickly sub-
sided. In a way, I suppose I was lucky to
learn so early that treasure diving is a risky,
frustrating business. I have since hunted
treasure only for fun, not profit.

As for my companions, Mendel moved
into marine archeology and Art remained
a professional gold seeker. Together, they
represent two camps of undersea treasure
hunters—those looking for historical

her companions. Her prize: a clay pot from a shipwreck 90 feet below. Clear waters here give divers good visibility.

treasures and those seeking fortunes. I have dived with both groups, and know well the archeologists' complaint that treasure divers rip apart wrecks, destroying irreplaceable artifacts. Fortunately, today's treasure hunters are increasingly aware of a wreck's archeological importance. Many began as gold seekers and later shifted to marine archeology. This signals, I think, a more productive future for treasure diving.

Gold, wherever it is found, can bring out the worst in men. But there's a lot of good to be said for gold seekers. Without their initiative, many wrecks would never have been discovered; the gold, the rare porcelain, and other beautiful artifacts now in museums would still be submerged. Were it not for the pioneering efforts of treasure hunters, the tools now basic to ocean work —airlifts, water jets, detectors, lifting gear— would not have evolved so quickly.

I've never regretted my farewell to treasure diving, but every now and then I recall a day when Kip Wagner and Mel Fisher showed me doubloons they'd found on a Florida beach. They were looking for the source of those coins and offered me a partnership in return for my time and the use of my salvage gear. But I knew that the vast majority of treasure seekers find less gold than my single doubloon! And so I turned them down. In time, they found their sought-for mother lode. In doing so, they joined the elite ranks of the very few successful treasure hunters.

But many others have found their own kind of treasure in the sea—the beauty of a coral reef, the sheen of a rare shell, or simply the joy of diving. This book is about all those who find treasure or pleasure beneath the waves.

EDWIN A. LINK

Contents

Waiting for evening fog to lift, a dive boat rides
smooth water near a wreck site off the Isles of Scilly
in the English Channel. Air bubbles and murky
water obscure a diver searching for Spanish gold
in the Straits of Florida.
   Treasures of the depths, from upper left: Door
latch, Confederate blockade-runner, off North Carolina.
Rare slit shell, Bahamas. Marble horse, first century
B.C. vessel, Greece. Oval gold medallion, Spanish, off
Northern Ireland. Earring, Spanish, Bermuda. Middle:
Monogram, French cannon, Isles of Scilly. Gold coins,
Spanish, off Florida. From upper right: Gold bar and
15½-foot chain, Portuguese, Bermuda. Marble statue,
Greece. Porridge bowl, Spanish, Padre Island, Texas.
Silver coin bearing the likeness of Charles VI, Holy
Roman Emperor, Shetland Islands, Scotland. Silver cob
struck in the New World, Spanish, Florida. Encrusted
signet ring, Dutch, Shetland Islands.

SHELL, DIVER, MEDALLION, SILVER COIN, SIGNET RING: N.G.S.
PHOTOGRAPHER BATES LITTLEHALES; BOAT, MONOGRAM: ROLAND
MORRIS; STATUES: GEORGE PETERSON, N.G.S. STAFF; GOLD COINS,
SILVER COB: N.G.S. PHOTOGRAPHER BRUCE DALE; GOLD BAR, CHAIN:
CHARLES ALLMON; LATCH: JOHN BROADWATER; EARRING: PETER
STACKPOLE; PORRIDGE BOWL: JOHN LOPINOT

Wrecks and Riches: the Lure Below

1

By Robert Sténuit

CURIOUSLY, my quest for sunken treasure began in a small bookshop. It was more than 20 years ago, on the Galerie de la Reine, Brussels, and I was 19, a university student of political science.

In the shop, I saw a paperback book, its cover illustrated with deep-sea divers wresting a chest of gold coins from a skeletal old shipwreck. I picked the book up and thumbed through its tales of sunken Spanish galleons, gold doubloons, and silver pieces of eight. Spellbound, I followed the

Cement-carrier *Caridad*, lost in 1970 in the Virgin Islands, intrigues an exploring scuba diver. In time, the sea reduces most shipwrecks to little more than bulges in the bottom sand.

author's schooner through each episode. Again and again he would navigate unerringly to the "X" scrawled on an ancient map, put on diving gear and jump, landing smack in front of the sought-for wreck. A cabin door would creak open, usually revealing a giant octopus, which the author swiftly dispatched. Then he brought up chest upon chest of gold and jewels, pausing only to rip open the belly of a menacing shark. It sounded very easy—and exciting!

Wasn't this "the good life?" Even at 19, I realized such stories couldn't be true, but why couldn't I make them happen? I knew how to use diving gear, I had no need to rush into a "serious" career, and the fires of ambition were lit. I bought the book. From that moment my political career was stillborn, my destiny set. I quit school to chase golden rainbows.

Childish? Perhaps. A dream? Definitely. But it was a dream that would become a life-style, luring me to all the oceans in search of gold-laden galleons and East Indiamen, 18th-century warships, and World War II booty. That very summer I set out on my first treasure hunt with a visit to Vigo Bay in northwest Spain, a spot my book recommended highly for sunken gold.

But first I visited libraries in Brussels and Paris to satisfy myself that a fleet of treasure-filled galleons had actually gone down in Vigo Bay in 1702.

It had.

On June 12, 1702, a Spanish fleet heavy with New World gold and silver set sail from Havana. I studied the extensive bills of lading, aware that much more wealth probably lay hidden in the officers' cabins. In those days Spanish nobles paid the king dearly for their commissions; a treasure fleet was their chance to recover the fee and make the discreet profit that would see them through old age. Surely they had salted away unrecorded fortunes in emeralds, Indian goldwork, amethysts, and rubies.

But these Vigo cargoes would buy no one's pension: An Anglo-Dutch squadron attacked the fleet in Vigo Bay, and every Spanish ship was sunk, beached, burned, or taken. Untold treasure went to the bottom. Although Spanish divers later salvaged much gold, most remained submerged.

In my youthful enthusiasm, I believed that modern technology—self-contained underwater breathing apparatus, or scuba—would assure me success where others had failed. Gleaning ship locations from archive records, a friend and I summered in Vigo Bay—and stumbled on three wrecks. But thick mud filled each hulk to the deck line, blocking our attempts at excavating by hand. We left Vigo Bay somewhat frustrated—we had found ships but no treasure.

Returning to Vigo Bay on a later expedition, I spent two months of backbreaking work with an airlift—a giant undersea vacuum cleaner—and finally bared one wreck. But I soon realized we had been working on only half a wreck, and the wrong half at that. We had uncovered her fore section, and any treasure she might have had would be in the aft holds. That part of the ship had broken off and was nowhere to be found. The "treasure" we recovered consisted solely of cannon, corroded daggers, tobacco leaves, and broken pots.

We then tried to locate another wreck, the richest of the fleet. It had sunk in 300 feet of water outside the mouth of Vigo Bay, thus eluding divers for centuries. Working from archive records and a knowledge of tides and winds of the early 1700's, we picked a search area and dived. Almost immediately I found a wreck with iron guns and cannonballs. That night we celebrated, but the next day we hauled up one of the cannon, and I spied an English ordnance mark that dated it to the late 18th century. We had found the wrong wreck!

Over the years I would find a number of
(Continued on page 16)

Exposed by treasure hunters, wooden ribs of the Spanish
galleon *San José* — lost off the Florida Keys in July 1733 — jut
from the sand that preserved them for nearly 2½ centuries.
Bound for Spain with gold and silver from New World mines,
the ship foundered in a hurricane with 21 others — most in
shallow water. Slaves diving for Spanish masters recovered
much of the cargoes, but the wrecks still yield artifacts and
coins: Finds from the *San José* alone total $500,000.

JERRY GREENBERG

Overleaf: Nearing the surface, a sport diver
steadies a ship's wheel brought up from 150
feet during a weekend dive on the *Varanger*, a
Norwegian tanker sunk off New Jersey by a
German submarine during World War II. An
antique dealer has since restored the wheel
and values it at $1,750. Amateur divers usually
haul such trophies home.

Riding a mountain of water, the dive boat *Sea Ranger* pitches in
Atlantic swells. Aboard, divers suit up for a descent on the Dutch
freighter *Arundo*—one of 110 ships torpedoed by U-boats along the
Eastern Seaboard; members of the diving club later brought up the
ship's bell. At left, a diver ascends with a lantern from the
Varanger, a lobster bag for collecting artifacts, and a flashlight for
exploring dark, crumbling interiors.

Coral-encrusted cannon mark the site of the French frigate *L'Herminie*, where diver Edward (Teddy) Tucker fans the sand for artifacts. In 20 years of treasure diving, Tucker has mined hundreds of ships on the wreck-strewn reefs around Bermuda. The Portuguese coin came from an unidentified wreck of the 1580's.

wrong wrecks: wooden and iron trawlers, 20th-century boiler plate, even—on several occasions—the telegraphic cable to Portugal. I would endure jobs as salvage diver and ship broker—anything to keep afloat financially and still satisfy my keen appetite for diving. Also, I would come to know the misconceptions and frustrations of treasure diving. Looking for a galleon with tattered rigging jutting from the sea floor? Forget it. Usually there is no recognizable wreck. Strain your eyes to see some skeletal hulk and you will probably miss that barely perceptible hump of a slightly reddish brown that is all that remains of a cannonball in the final stages of disintegration. Neglect that hump and you ignore the one visible clue to nearby treasure buried in the sands of endless tides.

Of course, if there is no visible wreck there can be no creaking cabin door, no treasure chest, no guardian octopus. Sharks? Treasure divers don't notice the indifferent sharks; they're too busy shifting pebbles or fanning away sand, much as a land archeologist sifts tons of overburden to uncover relics from the past. Compared with fiction, real treasure hunts can be terribly dull. They consist mainly of divers probing underwater with unromantic coal shovels or airlifts five or six punishing hours a day, month after month, year after year. Treasure divers endure muddy water, bad weather, tricky and dangerous currents, and often that cold, cold water.

Still, I enjoy it—more than anything else. The cold and the other problems are only a challenge. It's no fun to climb the easy slopes, is it? And I enjoy the physical labor, balanced by the intellectual research that is the basis of any successful dive. But best of all, each time I go down I am as thrilled as the first time I strapped on tanks—I become another animal, with new eyes and new limbs. I sense and feel different things when I enter that underwater fantasyland bathed in murky green.

Even during the disappointment of Vigo Bay I returned from each day's work exhausted and empty-handed but gloriously happy. For I was doing exactly what I wanted. I valued my new-found freedom and longed to continue diving, to make as few concessions as possible to the conventions of modern times. Those years in Vigo Bay were the best of my life.

Except for all the years since then. For in addition to the many intangible joys of diving, I'll never forget the thrill of finding my first sunken cache of gold, after 15 years of searching. The ship was the *Girona*, which sank in 1588 equally overloaded with treasure and history. A proud galeass—half galley, half galleon—the *Girona* sailed in the "Invincible Armada," some 130 ships sent by Philip II of Spain in a religious cause against Protestant England. Philip hoped, too, to end the pillaging of his treasure ships by Englishmen like Sir Francis Drake. In a Channel battle, the Spanish—armed with heavy cannon—used up all their shot without seriously damaging the English galleons. A change of wind blew the Spanish north. As the ships ran for Spain via Scotland and Ireland, autumn gales sank a number and threw 20 or 30 more onto beaches and cliffs. One was *Girona*.

For more than 600 hours, I dug in the dusty archives of five countries, seeking precise details on *Girona*'s last voyage and resting place. The galeass was off what is now Northern Ireland when its jury-rigged rudder gave out. Huge seas flung the ship sideways against crags that gored the hull,

then snapped it in two. At dawn, five survivors lay freezing and battered on the black rocks of Lacada Point.

The water was rough and icy when I arrived in 1967 with old friend Marc Jasinski, a photographer and diver. We dived anyway, and the hours of patient research paid off: Within an hour we found *Girona*'s grave, in 30 feet of water. Of the hull, not a splinter remained—only fragments of lead sheathing. But cannonballs and brass guns pointed the way to the first Armada wreck ever discovered. We soon found scattered pieces of eight, tarnished by centuries of immersion. I started digging in a small mound of sand, and there, gleaming before my eyes, lay a few links of gold chain, untouched by the corrosion that marked our other finds. I was neither shocked nor stunned at finding my first sunken treasure, but felt a peaceful sense of fulfillment. This victory meant I could continue diving.

Marc and I left the site but returned months later, in 1968 and 1969, financed by the National Geographic Society and my patron, industrialist Henri Delauze. Our team expanded to five men. We drew up a chart of the sea floor and plotted the location of each find. For ten months we hand-sifted thousands of cubic feet of gravelly sand down to bedrock, exploring every crevice for the ship's long-hidden treasure. We shifted multi-ton boulders with the help of buoyant, air-filled neoprene bags.

Our painstaking search paid off with coins—405 gold, 756 silver, and 115 copper. And we found evidence of man's vanity—gems and finely-chased gold that the Armada's nobility planned to wear at the hoped-for victory celebration in London. Cameos in lapis lazuli gazed placidly from settings of gold and pearls. Handfuls of exquisitely carved gold buttons and the

(Continued on page 24)

Intended for battle against England's navy, arquebus shot and
stone cannonballs pattern a beach at Port na Spaniagh in
Northern Ireland. More than 600 hours of research in Europe's
archives led Belgian diver Robert Sténuit to the Spanish
Armada ship *Girona*. She splintered on outlying rocks in
October 1588 while retreating around Ireland. Above, author
Sténuit wears a solid gold chain, one of eight discovered
during three summers of undersea excavation on the site. Other
prizes from the *Girona* include a silver dolphin with upturned
tail—perhaps a cup handle or the knob of a lid; a Neapolitan
ducat bearing the coat of arms of Charles V, Holy Roman
Emperor; gold buttons engraved with flowers; and gleaming
gold coins. Lured as much by history as by gold, Sténuit finds
the greatest rewards of treasure diving in the challenge of
archival detective work and the weariness after a day's dive.
The Ulster Museum in Belfast now exhibits *Girona*'s treasures.

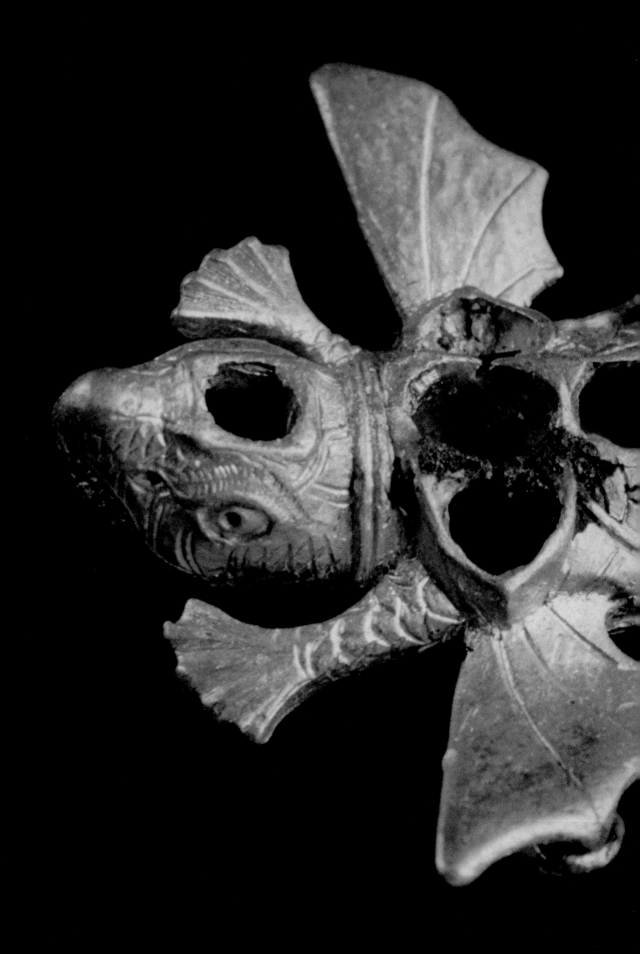

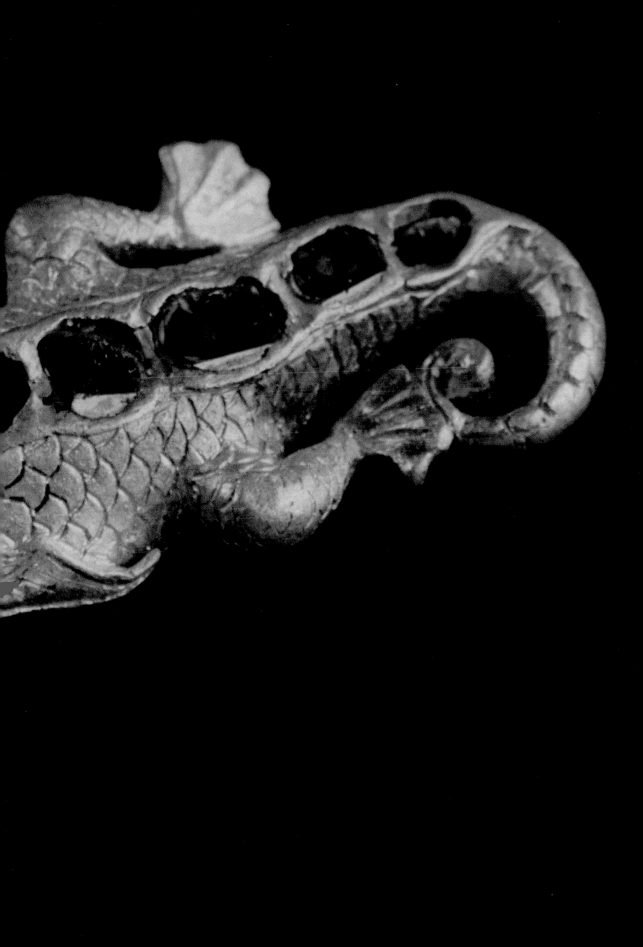

Set with pearls and a ruby, a gold ornament—perhaps a hat medallion or a link from
a chain—testifies to the wealth of a passenger aboard *Girona*. Many young aristocrats
in the Armada's ranks, already twice shipwrecked during the retreat, went down with
the ship. One, a Knight of Alcántara, wore a medallion of the chivalric order
(above). The map below, drawn by an English cartographer in 1588, shows ships of
the Spanish fleet sunk along the Irish coast. *Girona*, the last ship lost, does not appear.

Overleaf: Row of rubies once formed the spine of this winged salamander; only three
remain. Because legend accords salamanders the ability to live in fire, Sténuit
conjectures that a nobleman aboard *Girona* carried the ornament as a good-luck charm.

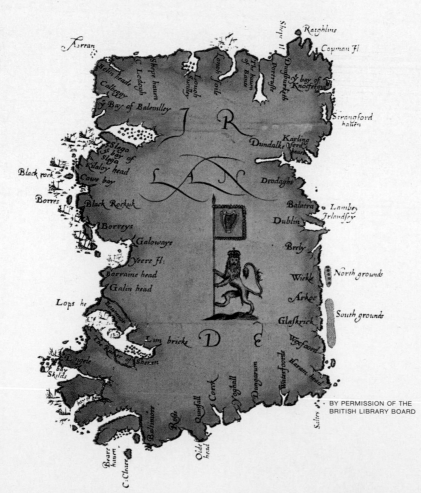

ROBERT STÉNUIT

cross of a Knight of St. John of Malta attested to swashbuckling nobles.

But my favorite find from the *Girona* is a slender ring engraved with a tiny hand offering a heart and the words *No tengo más que darte*—"I have no more to give thee." This delicate memento of 1588 reached across the centuries, conjuring in my imagination a beauty who gave it to her lover before he set sail. I felt a touching bond with the past I was exhuming.

Girona was not lost for lack of navigational aids—we found two astrolabes, five pairs of navigational dividers, and three sounding leads. Countless potsherds, pewter plates, copper vessels, daggers, and medallions gave us a glimpse of daily life aboard ship. Chandeliers meant nobility, as did gilt dinner services. I chuckled as I pictured the fop who owned the solid gold dolphin groomer that opened into a toothpick and ear-cleaning spoon.

Fantastic and exaggerated tales of our finds had reached shore, and one day 12 amateur pirates invaded our site. We five immediately took up positions, and luckily no knives were drawn. Some of the invaders dived, and I followed. I saw one slip something from the *Girona* into his bag. We had worked too long and hard to be deprived of our victory—I seized the bag. Suddenly the fight was on! Two pirates yanked my flippers off. I grabbed their crowbars, and we kicked each other all the way to the surface. A shouting match ensued, reinforcements rushed in for both sides, but the pirates had had enough.

Today, the treasure they hoped to take lies in its entirety in the Ulster Museum in Belfast, a fitting repository for the relics that had lain so long in Irish waters. As for the would-be pirates, I met them years later at a skin-divers' convention.

"No hard feelings," they offered, quite friendly on dry land. It was all an unfortunate misunderstanding, they said. I gladly accepted their outstretched hands. Rarely do competitors for undersea treasure reach such amicable accord.

I had run into far less friendly rivals in 1963, when I joined underwater explorer Edwin A. Link in Corsica. We had hopes of locating what is popularly but incorrectly called "Rommel's Treasure." The hoard, rumored to be gold and jewels stowed in six lead-sealed iron ammunition boxes, was a bit of World War II booty amassed by Gestapo Commander Heinrich Himmler from banks, museums, and private collections in the wake of the German armies. Field Marshal Erwin Rommel was not involved in any of these doings, but for some reason his name is associated with the treasure. It was submerged off Corsica at the war's end, outside French waters.

We arrived in Corsica with a copy of a map purported to have passed through concentration camps and prisons, where it had bought some of its owners freedom and others death. In the years after the sinking, several attempts to recover the treasure always ended in misfortune and intrigue. We had not been in Corsica long before we realized we were being watched....

An ominous letter—apparently from Corsican gangsters—warned: "Three men have died already in this affair. It is better for you that the list should not get any longer. The treasure belongs to us. This advice comes to you from well-wishers in Korsica and Nice."

Every day, two pale men in dark glasses, bright shirts, and light straw hats—city types, not fishermen—circled our ship in a speedboat, taking compass bearings. If we hadn't seen the letter, they would have been comical figures—a Marx brothers parody of Corsican gangsters. But late one night we heard their unlit boat approach. We scrambled to the deck. Was this an attack? We turned on our lights, heard the other boat pull away, and saw its wake. Relieved, we concluded our visitors had come not to maraud but only to make sure we were not diving at night. We soon learned to ignore their visits.

Ed Link had brought two magnetometers —devices invaluable for a treasure hunter. Towed across a body of water, they record the steady force of earth's magnetic field. Metal objects beneath them cause a slight change in the force, and the sensitive instruments measure the difference. With them we hoped to locate the ammunition cases on the sea floor. For days, then weeks, we towed one magnetometer along a grid marked by buoys, slowly charting each reading. At first our methodical search repaid us only with frustration and weariness.

And then it happened. The needles of the magnetometer jumped, went mad for a long moment, then slowly returned to normal—a textbook signature for iron or steel. We crisscrossed the spot: each time the signal reappeared. We had found a strong magnetic response in an area 400 feet in diameter. Should we dive on such imprecise data? I plunged in and found a Sahara of very soft sand. Burrowing my arm in up to the elbow, I felt no bedrock. Surely any heavy object would sink yards deep within a few months—and this was 20 *years* after the sinking. And if the boxes had decayed, their contents would be scattered through the ever-shifting sands.

Still, the fact remained that our long-range detector had done its job, so we switched to the smaller, more sensitive magnetometer to pinpoint the cache. But the device could not detect small quantities of iron buried more than six feet deep. Again and again we tried to find that treasure, without a trace of luck. The day came when Ed Link had to leave, with his equipment, and we returned home—empty-handed.

Had the first magnetometer really found the ammunition boxes, or just the engine of a lost airplane, or the wreck of some old

Etched flowers blossom along the stem of a silver spoon from the *Curaçao,* a warship that escorted a Dutch East Indies fleet in 1729. Sténuit found the remains off the Shetland Islands north of Scotland, where a current had swept her onto rocks in a fog.

ROBERT STÉNUIT

trawler? We may never know, at least until someone makes a better detector.

A decade after abandoning Corsica's fruitless sands, I was searching for older, more reliably pinpointed treasures. One was *Lastdrager,* a 17th-century transport, one of the Dutch East India Company's first ships to embark on the spice-drug-silk-porcelain trade to the Orient. I spent a winter in The Hague, combing records that line miles of archive shelves. There, from dry summaries of ship departures and cargoes, I chose the *Lastdrager.* Then in Amsterdam I found a "guidebook"—a manuscript written in 1653 by one of *Lastdrager*'s survivors, Johannes Camphuijs. He told of terrifying days and nights adrift in raging gales before breaking up on the rocks of Yell, one of the Shetland Islands. On board were some 25,000 guilders; only two chests of coins had been salvaged immediately after the wreck. The rest lay on the bottom—unless some treasure seeker had beaten me to the site.

I returned to the archives—this time in Scotland—for another month, until I was satisfied that only William Irvine, an 18th-century wrackman, or treasure hunter, had visited *Lastdrager.* He had gleaned just six silver coins, one medal, some lead pigs, and two anchors—mere crumbs of my treasure. I hurriedly secured exclusive rights to the wreck from the Netherlands government.

Arriving in Yell in 1971, again with Marc, I reread Camphuijs's painful account: "I was thrown on the rocks . . . I started to climb and crawl . . . I reached the cliff top . . . Naked, bleeding, covered by snow blown by an ice-cold wind, I started walking . . . Suddenly I fell in a freshwater well that was so deep I almost drowned. I resumed walking . . . I saw sparks flying in the sky. Walking towards where they came from, I found a stone house with a blacksmith's shop . . . I had survived."

I retraced the trail of this lonely survivor back from the stone house—in ruins now but still recognizable—to the well and the rocks where *Lastdrager* sank. There could be no doubt. The following day we dived and found the ship's scattered artillery just where Camphuijs had indicated it would be.

The site was in shallow water, but thickly blanketed with kelp. We could only crawl and pull ourselves through with our hands. But this undersea jungle was streaked with silver and gold! In the gullies and crevices of the sea floor lay hundreds of silver ducatoons and Spanish colonial pieces of eight, scattered by the action of swells. We found 400 small Dutch coins in piles, cemented by salts leached from copper in the coins. We took these black conglomerates to the surface where, once separated, the coins proved to be of vastly different mints and years. So, the Dutch East India Company had packed *Lastdrager*'s hold with existing moneys, not just coins from the Dutch mint, as the archives had recorded. Already my excavation had shown that all information—even that from official documents—should be treated with skepticism by a treasure seeker.

Apart from the 25,000 guilders, the ship records gave no indication of *Lastdrager*'s cargo, so we moved boulders and sieved the sand and gravel for clues. Artifacts—often amalgamated with natural concretions—were taken topside, cleaned, and studied, in order to learn as much as possible about our ghostly East Indiaman.

We found thousands of fragments of brass kettles and buckets, as well as copper sheathing, apparently to be traded for spices and other Oriental goods. One day I saw what appeared to be a liquid mirror on the sea floor—spilled mercury lay in rocky crannies, still shining like new. A syringe easily picked up the elusive quicksilver, needed in India to refine silver ore.

Hundreds of superbly crafted brass navigational dividers especially fascinated me. Some were so well lubricated with tallow that they opened easily, in spite of 300 years of immersion in the corrosive brine. Pocket brass sundials told an amusing tale of human foibles. Made in Germany, the timepieces were designed for that latitude and would be useless in China and Indonesia, where they were to be sold. But wealthy Orientals bought such instruments, apparently for the same reasons that European contemporaries acquired Oriental curiosities useless to them. They were pretty and would make nice conversation pieces.

Unfortunately, our dive was not without disappointment. It was soon obvious that, as in Vigo Bay, we had found only the fore half of the wreck, and thus were missing the bulk of the treasure and cargo. Wooden ships often broke up in storms, and the separate halves commonly came to rest yards or miles apart. According to Camphuijs's diary, *Lastdrager*'s stern had indeed drifted out with the turning tide. We followed its faint trail of mercury, cannonballs, and ceramic shards, but the spoor ran out in a sandy, trackless desert a hundred feet deep. We dived for days in a systematic search pattern before admitting defeat. Even with a

(Continued on page 34)

Overleaf: Dismasted and helpless, the Danish frigate *Wendela* breaks up on the rocks of Fetlar in the Shetland Islands. Bound for India in October 1737, she carried 79 silver ingots and 31 sacks of silver coins. Contemporary letters tell of "boysterous" weather and dismembered bodies flung high up the cliffs. Alerted by flotsam driven ashore, islanders came to comb the sea bed with grappling hooks and dragnets, bringing up most of the *Wendela*'s treasure. Sténuit began diving for the rest in 1972.

PAINTING BY JIM BUTCHER (OVERLEAF)

At Sténuit headquarters on Fetlar, diver Maurice Vidal pumps compressed air into a supply of tanks. In good weather, the men spent six hours a day in icy water systematically dislodging boulders and sifting sand, seeking coins from the *Wendela*. Each evening they repaired equipment, catalogued finds, and began cleaning and restoring artifacts. At right, the team departs for the dive site from Funzie Bay. Its rocky beach hampered portage of the boat, engine, and heavy diving gear, but it provided a calm spot along the exposed, wave-lashed coast. Clad warmly in brisk August weather (above), Sténuit carries charts for plotting the site of the *Wendela*.

Broken and eroded by the sea, a cannon from the *Wendela* hangs from a lifting balloon; divers moved it to scour the spot where it lay. Lacking facilities for preserving the iron guns, they left the 16 they found on the ocean floor. Sténuit's team also used balloons to move some 200 boulders. Only where they could expose the sea bed did they find treasure—coins and artifacts had slipped between rocks to the bottom. Among the forest of kelp that further hindered work, another diver inflates heavy flotation bags with compressed air. Hard to control in swells, the neoprene balloons contributed strained muscles and exasperation to the divers' difficulties.

magnetometer we found nothing but highly magnetic rocks. We could only conclude that powerful tides had carried the slowly bulging aft half of the ship far out to sea. We had recovered nearly 3,000 objects from the ship, but nowhere near the 25,000 guilders we had hoped for.

Still, I was excited and pleased by our finds, especially a few grams of greenish brass—a small jointed pointer badly corroded by the sea. It may look like nothing, but I'm immensely proud of it, for it is part of a Dutch mariner's universal astrolabe—one of only two fragments of such an instrument surviving today.

A few years later 18th-century treasure hunter William Irvine led me to another wreck in the Shetlands, the 26-gun Dutch East Indiaman *Wendela*. Carrying 79 bars of silver and 31 sacks of silver coins, she broke up on the isle of Fetlar in 1737. According to documents I found in the Scottish Record Office, Irvine heard of the wreck early in 1738. "There are lately driven in several dead men's bodies and some masts and yards," he wrote to the Admiral of Shetland. "By a Danish book that is saved it is probable that she is a Dane. . . . she has sunk in a very barbarous place."

Irvine rushed to the site with his gear and his "diving machine," a weighted barrel with crystal portholes and openings for a diver's arms, but local farmers salvaged most of the treasure. With nets and grap-

During a day ashore, Louis Gorsse of the Sténuit team repairs one of hundreds of brass dividers taken as cargo aboard the *Lastdrager*, a Dutch transport claimed by the Shetlands in 1653. Below, a pair of different design— probably used by *Lastdrager*'s navigator—rests on a chart showing Bluemull Sound, where she sank. Gold and silver coins from *Wendela* provided knowledge about the currency of the period: The 717 recovered represent a diversity of countries, values, mints, and dates.

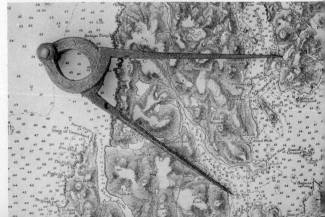

pling hooks they fished up 60 silver bars and the contents of at least 22 sacks of coins.

Even though there wasn't much treasure left, we dived on the site and recovered 44 gold coins—nowhere mentioned in the official bills of lading—and 673 silver ones.

By the time I found *Lastdrager* and *Wendela,* my research methods had become routine. I first would locate a wreck as certainly as possible on paper, then—and only then—set out to find it in the sea, knowing as much about that single vessel as any library or archives could tell me. But one day I found a wreck quite by accident while on the trail of ships entirely unrelated to it. I had to reverse my usual methodology and fit the artifacts from the dig to archive records.

We were diving for a fleet of Dutch West Indiamen that had run onto reefs off the isle of Whalsay in the Shetlands. But the ship we found did not belong to this fleet. It yielded 18 iron cannon and various cannonballs of the late 18th century. In newspapers of the time I found accounts of a Russian wreck, so I asked friends of friends to check the archives of the Russian Imperial Navy. They found one possibility—the *Evstafii,* a Russian frigate that had sunk here in 1780. But she carried 36 guns, not 18. Had we found a smaller ship than *Evstafii*? Or had the *Evstafii* jettisoned half her cannon before running aground?

Another look at the records revealed that the laird of Whalsay had rescued the *Evstafii*'s survivors, salvaged some masts, sails, barrels of pitch, and other miscellaneous debris—and 18 iron cannon. That explained the missing armament.

We completed our excavation of the wreck, and were soon positive of the ship's identity. Small arms bore Russian initials. Big copper kopecks littered the site, lost probably from the pockets of sailors. We also turned up some shining gold ten-ruble coins with the portrait of Empress Catherine II—the proper currency for a 1780 wreck.

Nearby, we unearthed a beautiful collection of icons—St. Nicholas flanked by Christ and the Virgin Mary. Here was proof of the wreck's nationality—and of the wreck itself, for no other Russian ships of this era had gone down near Whalsay. I was pleased with my success, but I also felt a twinge of melancholy whenever I looked at those icons and wondered about the terrified mariner who had clutched one in his hands in his final hours.

You see, I've grown up a bit from my gold-fever days. Ever since Vigo Bay my thirst for gold has gradually decreased and made room for the other joys of treasure diving—the freedom, the allure of fishlike movement, the excitement of unearthing artifacts that haven't seen daylight or dry land for centuries.

I now find myself less interested in the account books of a silver fleet than in what sort of person wore the gold ring that is today the only tangible record of his existence, his life, and his loves. I have metamorphosed from gold-seeker to amateur marine archeologist.

A wreck is not just a pile of debris to be plundered. It is an old lady deserving of the ultimate respect from divers, for she is the sole record of a certain voyage in a certain age of history.

Today I still seek—and respect—those grande dames of history. My winters pass in archives, where like a detective I unearth the "facts" about my quarry. Summers are spent blissfully diving for the wrecks I've researched. With me, treasure diving is not just a hobby or even an occupation; it is a way of life. And I am very happy.

I suppose I could have been successful in a more commonplace occupation. But I'd have wasted a good part of my life. And why? I'm rich in what I value—good memories. My only regret is that I waited so long before hunting down *Girona.* Why didn't I start chasing my rainbows *sooner*?

"A very barbarous place," a Scottish official called the dark cliffs of Wimligill Burn after overseeing the first salvage operations on the *Wendela* in 1738. The author surveys the ship's turbulent grave 30 yards offshore. Numerous and explicit references in old records to the promontory—"the huge rock where she is wracked"—pinpointed *Wendela*'s remains for Sténuit; his team found them on the second day of diving. On hydrographic charts (below), Fetlar farmer Jamesie Laurison points out to Sténuit locations established as wreck sites by island tradition. He often finds such local lore more reliable than official documents or maps.

A treasure hunter, he believes, "should be more inquisitive than acquisitive," for the search requires patience, persistence, and hard work. The coins and artifacts retrieved, while often historically important, seldom repay the cost of finding them. Even so, the excitement of a strike can spark dreams of gold, and whet the appetite for more.

NATIONAL GEOGRAPHIC PHOTOGRAPHER BATES LITTLEHALES

Pioneers in the Search for Gold

2

By Mendel Peterson

THE green fire dazzled me. Seven emeralds, set in a softly glowing gold cross, burned with an ethereal light. I knew at once that I was holding one of the most valuable pieces of treasure ever taken from the ocean. A young Bermudian named Edward (Teddy) Tucker had found this exquisite 16th-century cross near a reef beneath the clear waters northwest of his island home.

For me, the story of the cross began a week before I saw it, early in November of 1955. As Curator of Naval History at the Smithsonian Institution in Washington,

"The sea's been a generous lady to me," says Teddy Tucker, Bermuda's most successful salvor, here aboard his workboat, "but many divers find more frustration than treasure."

D. C., I was asked to examine some color transparencies of Spanish artifacts and treasure recently recovered by Tucker. Here we go again, I thought, another false alarm. Treasure claims and fictions had been popping up regularly for years. But as I looked through the photographs, I realized that Teddy had made the most spectacular haul from under the sea since 1687, when an American, William Phips, salvaged a little gold and 26 tons of silver from reefs north of Hispaniola.

A week later I arrived in Bermuda eager for a firsthand look. That night must surely be compared in its own way with the opening of a Pharaonic tomb in Egypt.

On a coffee table was a dazzling array, not only the stunning cross, but gold ingots, round like biscuits, each weighing almost two pounds . . . a gold bar a foot long weighing 2½ pounds . . . two small sections from a similar bar which had been chiseled off to "make change" . . . delicate gold filigree buttons set with pink conch pearls . . . a mass of silver coins, blackened from the action of chemicals in the sea water. Most were pieces of eight, once the standard international currency of the world and familiar to every reader of *Treasure Island*. All of the artifacts and coins indicated that the ship had sunk around 1595.

For years, Teddy had been salvaging metals from the many shipwrecks around Bermuda, using the diving skills he had learned in the British Navy. He concentrated on the modern ships containing large quantities of brass and copper, which brought good prices as scrap. Combing the reefs in a small open boat with his brother-in-law, Bob Canton, was a chancy business, but they made a living and avoided the tedium of regular jobs.

In the summer of 1951, as they were returning with a load of metal to home port at Cavello Bay, Teddy spotted some long objects lying far out on the reef flats to the northwest of the islands. They anchored and Teddy went over the side.

"They were iron cannon encased in the sand plaster of ages," he told me. "There was also a large copper pot filled with lead musket balls. I knew it was an early shipwreck site although I'd never seen anything like it before." He sold the guns and pot to the Bermuda Monuments Trust, a private organization of volunteers, and filed away the position of the wreck in his mind.

Four years later he returned and recovered the treasure I saw on the coffee table. In 1961, after some negotiating, Teddy sold the treasure, historical artifacts, and the museum in which he had displayed them to his government for $100,000, to keep the collection in Bermuda; the emerald cross alone was worth much more. Every season since, Teddy has brought up more treasures from the reefs around Bermuda.

I dived with Teddy on the site where he had recovered the cross two summers after the great find. Plunging to the bottom, I found myself on the flat, sandy floor of a small amphitheater. A heap of ballast stones trailed off toward coral walls alive with swaying plants. At the foot of a coral cliff, Teddy pointed out the spot where he had found the treasure. We fanned the sand with our hands to see what might turn up. Nothing. I examined the flint ballast stones. Even if Teddy had not found dated coins, the ballast would have pointed to about 1600 as the latest year the ship could have gone down, for after the invention of the flintlock gun and of fire starting devices, flint became too valuable to use as ballast.

During the three weeks I spent with Teddy we found another small cannon, nails, tools, and fragments of timber. Even more important, I found Teddy to be an intelligent, skillful observer. The hours together underwater began a long friendship that led to our annual summer archeological dives for the Smithsonian Institution.

Twisted wreckage of the English steamer *Caraquet*, lost off
Bermuda in 1923, yields Teddy Tucker a massive bronze
propeller blade and other valuable scrap metal. Two years
later, in 1955, Tucker stumbled onto a 16th-century galleon, a
find that transformed him from scrap salvor to treasure hunter.

Teddy was not the first treasure diver I had known. In 1951, I met Arthur McKee, Jr., the first modern diver to explore the remains of a Spanish treasure ship in the New World. I had gone to Florida on my first diving expedition with some friends to look at a wreck that had earlier yielded some artifacts. Art joined our party, and one evening a few days later, after he had decided I could be trusted, he came to my room. With him was his diving partner, Wes Bradley. Art carefully closed the door and window blinds—my first taste of the fanatic secrecy with which treasure hunters guard their sites and their finds. From his pocket he pulled a small gold coin. I recognized it as a two-escudo piece—or quarter doubloon—struck at Mexico City in 1732.

In his matter-of-fact, confident way Art told me his story. As a teen-ager in New Jersey, he said, he "sneaked" his first dive one morning in 1934 when his boss, an old hard-hat diver, was too drunk to work. Art borrowed his suit and jumped in. Three years later, a trained diver, Art became recreation director for the city of Homestead, Florida, just at the head of the Keys. During his spare time he collected valuable scrap metals from modern shipwrecks.

"I saw my first treasure in the sea in 1937 when a man who had a chart showing the location of a wreck hired me to dive for ten days," Art said. "When I went down I expected to see a ship sitting there. All I found were some cannon." His disappointment turned to joy the next day when he went over the edge of the reef into deeper water and found the remains of a barrel. Beside it was a mass of tar with gold coins sticking out. "That mass eventually yielded 1,600 gold doubloons," Art said. The job finished, Art left with his wages, a few coins as mementos of his adventure, and an incurable infection of gold fever. It led him to a lifetime of undersea treasure hunting.

His first big find came in 1948. "My friend Charles Brookfield told me of some silver coins found by fishermen on Gorda Cay in the northern Bahamas," Art said. "Some friends and I went to the place, searched, and found nothing. But the next year we found an encrusted bar. I thought it was iron.

"On deck, when I struck the bar with a hammer, a piece of crust came off, and we were looking at bright, shiny silver. It weighed more than 70 pounds.

"I couldn't get back to the bottom fast enough," said Art, "for I had seen two more bars there." The bars and a few coins were all he found, but they were the first substantial Spanish treasure to come out of the sea since the days of William Phips.

Art later found several wrecks of the 1733 silver fleet disaster. On Friday, July 13 of that year, 22 Spanish ships, most of them loaded with tons of treasure, left Havana bound for Spain. The next day a storm, striking between Key Biscayne and Vaca Key, wrecked the fleet. A fortune lay scattered in 30 feet of water. Immediately a relief fleet located the wrecks and salvage operations began. But the divers had only primitive diving bells, crude grappling tools, and their own lungs. Salvors reported to the king that they had recovered all the treasure—and much contraband—but worthwhile quantities were left.

In 1949 Captain Reggie Roberts, a Florida fisherman, took Art to one of the sites, where guns plastered with coral sand lay about on a great pile of ballast stones. Art worked there off and on for the next 11 years, recovering pieces of eight and pieces of four, gold jewelry, worked silver, and a fine collection of artifacts. He had already brought up much of this in 1951 when he told me about it and showed me the gold two-escudo piece.

He invited Ed Link—inventor of the Link Trainer, an apparatus that helped thousands of World War II pilots learn to fly—his wife, Marion, and me to join him on

NATIONAL GEOGRAPHIC PHOTOGRAPHER EMORY KRISTOF
ACTUAL SIZE: 2 3/4" × 1 5/8"

his wreck. Art thus became my first underwater teacher. I came to respect his practical knowledge of diving and recovery techniques. We worked for two years on the site and recovered numerous artifacts; Art gave many of them to the Smithsonian. The winter following our first work on the site I located a copy of an old map in the Library of Congress in Washington. Apparently prepared by a Spanish navigator while the remains of the ships were still visible, it showed the locations of the wrecks of the 1733 fleet. Art's site corresponded with that of the *Rui*, flagship of the admiral.

Art's treasure hunting involved him in the kinds of adventures one expects to read about only in novels. Underwater poachers attacked him, for example, after he had worked the *Rui* for many years. Under an agreement with the State of Florida, Art thought he would have official protection of his site. But in 1960, when he learned that a group of men was coming down from Miami intent on looting his wreck, he was shocked to find that the state could do nothing. Art was driven off his site at rifle point and the intruders plundered the *Rui* and several other wrecks.

In almost 40 years of diving for treasure Art has visited 300 sites in the Keys, the Caribbean, and the Bahamas, and expended countless hours of toil as well as a substantial treasure of his own on the quest.

Art's success in finding treasure and the nationwide publicity he received sparked a rash of searches by divers in the Florida Keys. As the map and facts about the fleet of 1733 became known, more sites were dis-

Prize of a Spanish galleon sunk around 1595, this emerald-studded gold cross ranks as one of the most valuable single items ever recovered from the sea. Tucker plucked it from a Bermuda reef, then—to keep his treasure collection intact—sold it along with gold bars, coins, and other relics to the island's government.

covered. But most divers were poorly equipped and their success limited. Nevertheless, during the years between Art's discoveries and the first laws in 1958 requiring a lease to work wrecks in Florida waters, many antiquities and some treasure were recovered. Most were quickly scattered before they could be properly studied.

There was one exception. Craig Hamilton, a Miami fireman, permitted the Smithsonian to study artifacts from a wreck of the 1733 fleet—ship fittings, navigation instruments, tools, pottery fragments, coins of gold and silver, worked silver, and gold jewelry. A collection of this sort is precisely datable, with everything relating to everything else, and is of great importance to historians.

Just as Craig Hamilton was a fireman and Art McKee a recreation director, Kip Wagner, one of the few men to make treasure hunting really pay, depended for a long time on house building for a living. Like many of his neighbors in Sebastian, Florida, Kip had a vague notion that treasure ships had gone down off the beaches near his home. An occasional piece of eight was turned up by the surf after a storm. Anchors and guns had been found off the coast and some were displayed at Fort Pierce. Spending every spare moment during the late 1940's and through the '50's combing the beach, Kip collected hundreds of coins. When a friend told him that a 1715 Spanish fleet had sunk off those very beaches, he became a confirmed treasure hunter.

In July 1715 when the Spanish treasure fleet gathered at Havana to sail for home, the 11 ships carried tons of wealth which had been accumulating at Veracruz, Mexico, and Cartagena, Colombia, during several years of the War of Spanish Succession with the English and Dutch. Now, with the war over, New World gold, silver, and pearls, along with goods of the Orient that had arrived via the trade route from the Philippines and across Mexico, could once more be sent safely to Spain without fear of interference from ships of unfriendly nations. But just south of Cape Canaveral, a hurricane dashed the fleet on the sandy coast and all the ships but one were lost.

Kip, researching the story in a local library, found a copy of Bernard Romans' 1775 history of Florida and in it a map with an inscription indicating the location of the wrecked 1715 fleet. That, and microfilmed records from the Spanish Archives in Seville, convinced Kip that the wrecks lay off "his" beach. He began to search for the campsites of the 1,500 survivors and the Spanish salvors. When he found large depressions and mounds in one area of the palmetto scrub, he spent months walking back and forth across them with his $15 Army surplus mine detector, listening for the high-pitched whine that indicated metal. He found a rusty automobile spring and a cast-iron coffee grinder.

But finally, when he turned up a large wrought-iron nail and a cannonball, he knew he had found a camp. For days he dug in a half-acre site but found only broken pottery, cannonballs, two rusty cutlasses, and fragments of silver.

Then he saw what he had dreamed of for so long—the glint of gold at the edge of a hole. He lifted out a gold ring with a crudely cut pyramid-shaped diamond of about three carats. Kip now felt certain that at

Golden bounty from an unidentified mystery ship capped 12 years of searching by Bermudian Harry Cox. His haul—worth some $300,000—includes a 3½-inch-long gold manicure set (top) and gold circlets (bottom) carried as a convenient form of cash. In 1968, the year Cox began working the wreck, he found the gold ring; two years later he brought up an emerald that fit it perfectly. Other finds: an encrusted copper ring and a buckle.

CHARLES ALLMON

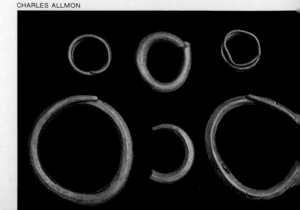

least one ship lay just off this part of the beach. Pelicans overhead gave him the idea for spotting the wreck—by looking down from a low-flying plane.

Soon, Kip was hanging by his safety belt from the open door of a small aircraft, throttled down to its slowest speed. He saw a dark blur on the gray sand bottom, along with objects like logs lying about—ballast stones and cannon. The next day Kip and his pilot were swimming on the site, looking into the water through face masks. "I had a hard time matching what I saw underwater with my imagination's image of a treasure wreck," he wrote later. "No timbers remained—probably all having long since passed through the digestive tracts of . . . shipworms. Then I swam over my first gun. It loomed larger than its original bulk. The sea had wrapped it in a limy crust embedded with shell fragments and plumed it with streamers of bright green seaweed that fanned with each surge of current."

The time had come to organize the first large treasure-diving operation, called the Real "8" Co., Inc. Wagner bought a surplus 40-foot U. S. Navy launch, christened it *Sampan,* and recruited a team of men interested in putting in time and money—his family physician, two colonels from a nearby airbase, a former Navy underwater demolition diver, two boatmen, and a banker who became business adviser. Later, two more military men, both divers from the Cape Canaveral Missile Test Range, joined the group. A lease issued by the State of Florida allowed them to work the site.

Equipped with shallow-water diving masks and air hoses from the boat above, baskets for lifting finds, buoys for marking the site, containers for holding artifacts, a boarding ladder, and scuba, they went for the treasure in April 1960.

"Putting out to sea in the *Sampan* during our first diving excursions always reminded me a little of the 'Owl and the Pussycat,' "

Kip said. "We had no cabin, so we nailed a tentlike canopy of sailcloth on a four-poster arrangement of two-by-fours. Our diesel engine chugged mightily to maintain headway in the rolling swells at Sebastian Inlet." Plunging overboard to the wreckage 20 feet below, the divers felt the same letdown Kip had when he first viewed the site.

They found only an indistinct mass of ballast and the encrusted guns. Of the ship's timbers and treasures, there was no visible evidence. Only fragments of the bottom of the ship, protected from the worms by the ballast stones, would survive for 2½ centuries. Heavy metal would sink into the sand. Silver pieces lying alone would be converted into silver sulphide by electrolytic currents. Coins in chests and bags would survive in masses. Coins of gold, "the eternal metal," would suffer no damage, and gold jewelry, if deep in the bottom sands away from the battering of the waves, would keep its beauty intact through the centuries. Iron fittings, tools, and cannon would acquire an encrustation of sand and calcium where plants and coral polyps would take root and grow.

From this camouflaged wreck Kip's team learned how difficult it is to retrieve what Neptune has claimed. Small holes fanned in the sand around the pile of ballast filled up quickly; the surging water made some divers seasick; sand stirred up by the divers and the wave motion cut visibility; moving ballast stones was exhausting.

But Neptune is a clever tempter—he permitted the divers to find thousands of fragments of red and tan pottery produced in Spanish America. Then in August 1960, Air Force Lt. Col. Harry Cannon picked up a small, strangely shaped, sand-encrusted lump. Cannon scraped the crust away and found a six-pound wedge of silver.

When the divers realized what they had, their hunting fever went up several degrees.

(Continued on page 56)

Crew of the salvage ship *Artiglio* sets out in 1929 to find the English liner *Egypt,* sunk in 400 feet of water off Brittany while carrying five tons of gold and ten tons of silver to India. Expedition leader Giovanni Quaglia (with cigar) searched with a dowser and divining rod before turning to more reliable draglines. Diver Alberto Gianni (opposite) designed a one-man observation turret for himself and other divers (left, with Quaglia); they would direct grapnel-wielding surface workers. But all three divers—and nine of the crew—perished while working another wreck. Quaglia reached the *Egypt* with a new team and fought six years of bad weather and delays before recovering nearly all the treasure—for Lloyd's of London.

GEOGRAPHIC ART (BELOW); PAINTINGS BY JIM BUTCHER

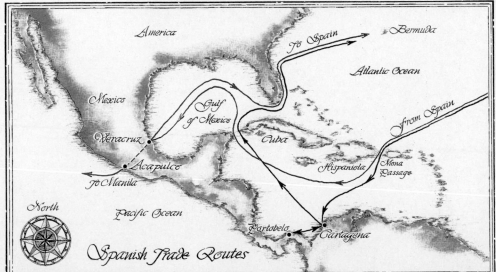

America

To Spain • *Bermuda*

Atlantic Ocean

Mexico

Gulf of Mexico

From Spain

Veracruz

Cuba

Acapulco

Hispaniola *Mona Passage*

To Manila

North

Pacific Ocean

Portobelo *Cartagena*

Spanish Trade Routes

Threatening to abandon mutinous sailors, treasure hunter William Phips quells a rebellion in 1684. Phips' failure to find sunken gold sparked this attempted mutiny, the second on a voyage that gleaned only trickles of treasure from Spain's wreck-littered trade routes (map). Spanish fleets called at two ports in the New World. One (its route in brown) sailed directly to Portobelo, where it traded goods for Peruvian silver, then wintered in Cartagena. The other (in blue) followed trade winds to Vera-cruz and picked up Oriental cargoes—tea, spices, and silks —brought overland from Acapulco. The fleets returned together past Bermuda to Spain.

Phips later stumbled upon the *Concepción*, flagship of the Spanish silver fleet of 1641. Clutching stones to speed their descent, Indian divers worked the 40-foot-deep site for six weeks; they distrusted the suspended, air-filled diving bell (below) and refused to use it. *Concepción* yielded Phips' sovereign—King James II—a little gold and 26 tons of silver.

Poachers with spearguns loom menacingly above Art McKee, intent on pillaging a Spanish galleon he has spent 11 years excavating off the Florida coast. Outnumbered five to one, McKee frightened off the attackers with a bangstick, a shotgun-like weapon he made to protect himself from aggressive sharks. One shot from the bangstick shattered a ship's timber, and the pirates fled—unaware that the gun held just one shell. But when the water cleared, McKee saw that the poachers had stolen two cannon the night before.

A decade later, more poachers plague McKee. One night in 1969 they riddled the vault in his Florida sunken treasure museum with 55 holes. But the door held and the would-be thieves left McKee both his gold and a fitting message (below).

Often called the "grandfather of treasure hunters," McKee has salvaged gold and silver from the waters of the Florida Keys, the Caribbean, and the Bahamas since 1937.

ROBERT MARX; PAINTING BY JIM BUTCHER

Drawn like sharks to blood, 300 gold-hungry raiders surround a surprised Spanish garrison in 1716. Their goal: bullion retrieved from the hurricane-wrecked Spanish silver fleet of the year before. British privateer Henry Jennings captained this bold venture. Quickly scattering the defenders, his men made off with some 350,000 pieces of eight before seeking refuge in Port Royal, Jamaica—then pirate capital of the world.

Many of Jennings' men later shed their semi-respectable cloak of "privateer," became out-and-out pirates—and died on the gallows. Jennings himself plundered more Spanish treasure, but quit when offered a royal pardon. Renouncing his past, he retired to a life of ease and respectability in Bermuda.

As for the Spaniards' sunken gold, divers continued to work the wrecks for several years, recovering millions of dollars in gold and silver—less than half of what officially went down. The remains of the hoard kindled 20th-century expeditions by treasure hunters Kip Wagner and Mel Fisher.

Progressing from beachcomber to millionaire, the late Kip Wagner first strolled Florida's sands for sea shells, then gold. Countless walks took him to Sebastian Inlet, where the surf occasionally turned up old Spanish coins. But only after moving offshore did he find their source—shattered hulks of the 1715 fleet. Some ships had broken up so close to land that Wagner's workboat—moored above them—lay a scant 1,100 feet from shore (left).

Eat-and-run diver Mel Fisher (right) joined Wagner's Real "8" Co., Inc.—named for *ocho reale*, the piece of eight—in 1963. Together they worked several 1715 wrecks, raising millions of dollars in gold and silver ingots, jewelry, K'ang-hsi porcelain, and coins by the bushel basketful, some worth more than $12,000 apiece.

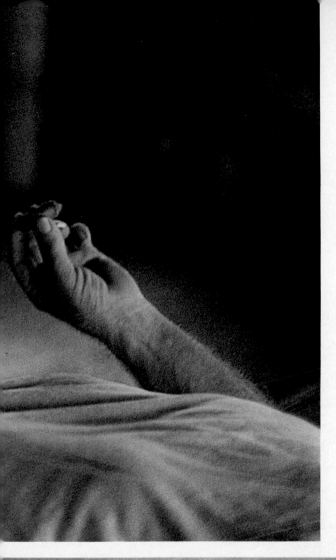

NATIONAL GEOGRAPHIC PHOTOGRAPHER BRUCE DALE

Nine more wedges were found. One now reposes in the Smithsonian, where thousands of visitors see it every year.

Through the 17th century and into the 18th, this odd wedge form was used for bullion, but no one knows for certain why. Early accounts mention silver wedges carried in baskets by mules and dumped loose in the square of Portobelo, Panama. Possibly they were packed in kegs like slices of pie before being loaded aboard the ships.

With a water lift, a flexible hose with water pumped through it, divers dug a trench around the pile of ballast stones, and spent months moving the stones into it.

On a memorable day in January 1961—the only day that month clear enough for diving—the men finally hit the mother lode. The ocean floor was covered with free and loose silver coins. "We were picking them up by the bushel basketful," Cannon said. None had a date later than 1715, assurance that the wreck was from the 1715 fleet.

After that one clear day, bad weather kept them off the site until spring. In April they were back, finding coins, plate for the table, cups, and other artifacts.

The next year, unusual ocean currents began to erode the entire east coast of the United States, and the divers were unable to do any work. Realizing that idleness can wreck morale, Kip organized hunting expeditions along the beach. On a ridge near their campsite, Kip's nephew saw a glitter-

Masterpiece of the goldsmith's art, this necklace and dragon pendant survived 2½ centuries of tossing surf and abrasive sand before Kip Wagner's nephew found it, undamaged, in a Florida sandbank. Tiny flowers emblazon each of the 2,176 links on the 11-foot-long chain. The dragon—a grooming tool—conceals a fold-out toothpick; the tail forms an ear-cleaning spoon, the mouth and body a whistle. In 1967 the relic sold at auction for $50,000— Wagner's richest single find.

ROBERT S. CRANDALL

ing piece of gold chain protruding from a sandbank. When he had finally pulled it all out he was holding the most valuable object Kip and his group ever found. The chain was miraculously wrought, each of the 2,176 links ornamented with tiny flower-shaped disks. Suspended from the chain was a golden dragon with ruby eyes. Its mouth and body were a whistle, its tail an ear-cleaning spoon, and a toothpick unfolded from the belly. Such a grooming tool would have been worn about the neck of a Spanish gentleman of the time.

The next season—the summer of 1963—Kip and his divers found their first gold on the wreck site: ten doubloons struck at the Mexico City mint. They also found gold jewelry and more silver.

In late summer a series of hurricanes raked the Straits of Florida and made working conditions impossible. Surging waves swept a blanket of sand over the site.

But in 1964 the sea yielded spectacular gold treasure. Another diver, former chicken farmer Mel Fisher, had joined the Wagner group and brought along a well-equipped vessel with an experienced salvage crew. Searching with a proton magnetometer—a device that incorporated recent advances in physics and electronic miniaturization—Fisher and Wagner located a new site off a beach where a schoolteacher named Frank Allen had found gold coins. Behind the propeller of the anchored ship Mel lowered his latest innovation: a large metal pipe bent downward at right angles. He called it his "mailbox" for its resemblance to a letter chute. Through the pipe the propeller directed a strong blast of water that in minutes washed away sand covering the site. This device gives the underwater archeologist nightmares, for it can be destructive if the blast is too strong. But used with caution, it quickly exposes the layers of "pay dirt" which can then be carefully worked using more precise methods.

Sand blown away, the divers plunged over the side. They worked several wrecks in April and May, found nothing, and were about to give up. "Then one million-dollar day in May," a diver told me, "I landed right in the middle of a solid gold circle eight or ten feet in diameter." In a few hours they scooped up a thousand gold coins.

Within a few weeks the site had yielded millions of dollars' worth of treasure. Thousands of gold coins that had spilled from a chest carried aft near the officers' cabins proved to be samples of all the denominations struck in the mints of Lima, Peru, and Mexico City, plus a few from Santa Fe de Bogotá, Colombia. For me, a numismatist, seeing the find was thrilling, for it contained many previously unknown varieties and increased our knowledge of that coinage a hundredfold.

With this find, Wagner was finally content. "I have reached my goal," he told a friend, "I am now a millionaire." Unfortunately, Kip died in 1972, but in the annals of treasure hunting, his success must go down as one of the greatest.

His group recovered 6½ million dollars' worth of treasure. A quarter went to the state, a portion to Fisher. Much of the rest, in the company's Museum of Sunken Treasure at Cape Canaveral, brings in thousands of visitors' fees yearly. Beginning in 1967, the company ventured into ocean survey equipment and engineering services; when income and a trickle of new treasure failed to pay the bills, they worked a year under limited bankruptcy. By mid-1974, the group had reorganized its affairs and was back to normal.

The story of another of the pioneer treasure divers, engineer Tom Gurr, has not turned out as well as Wagner's. But it illustrates the truism that many treasure salvors put $10 or $20 into the sea for every dollar a few take out.

If success could be measured by how hard

a man worked, Tom Gurr would be rich. For years, Gurr and some of his friends invested money, time, and sweat only to be disappointed. Still he kept his diving team together and finally, in 1968, settled on a promising site off the central Florida Keys.

Gurr believed his wreck was the *San José* of the 1733 fleet. But it was known that the Spanish had recovered much from the ship, and Gurr could only hope that enough treasure and artifacts remained to make the salvaging effort worthwhile for him and his financial backers.

Gurr and his divers first had to cut through a foot of matted sea grass roots to get to the sand covering the ballast stones. Laboriously, they removed the grass rug and the sand. A few weeks later I joined Gurr and his crew to look at some cannonballs, shards, and coins, and at the ship's timbers—almost all had survived, though flattened and scattered. I concluded that it was the *San José* without question. At my request, the timbers, ballast, and large objects were left undisturbed until I could return later to chart them.

Before Gurr started work on the *San José*, the Coast Guard had determined that the site was more than three miles from the coast of Florida. Presumably, then, it lay in international waters, and Gurr didn't need a state license. For six months the men labored, adding to their few gold rings and silver coins a collection of ceramics, glass, personal possessions of the passengers and crew, ship fittings, tools, and weapons.

Then, a rude shock. The State of Florida took Gurr to court and, after two trials, secured a ruling that Florida waters extended from the edge of the reefs—not from the land coast—and included Gurr's site. With no funds for an appeal, Gurr had to secure a state license that specified a 50-50 division instead of the customary 75-25 favoring the finder. The Federal Government won a court suit challenging Florida's contention but Gurr did not benefit; the case has been appealed to the Supreme Court, but a decision may not come until 1975.

On the site of the *San José*, Gurr and his divers by 1973 had brought up hundreds of thousands of dollars' worth of material. Some of it went to the state to be held until a division could be made, in accordance with terms of the licensing agreement.

On New Year's Day 1974, film of Gurr sitting in a small boat and dropping objects into the water was shown on network television. He said he was returning the treasures to the sea because he was bankrupt and the state was still holding the *San José* collection after five years. Many charges and countercharges later, Gurr and the state negotiated a settlement: He admitted he illegally sold some treasure, but had dropped none of it into the sea. He agreed to help retrieve it, and the state put him on probation while dismissing all other charges connected with the *San José*. Whatever Gurr receives of the treasure, by the time he pays his lawyers, stockholders, and crew, the chances are he'll be left with nothing.

Gurr's treasure hunting experience has been more dramatic than most—but I have yet to hear of a big treasure find that didn't end up in some kind of legal wrestling. Treasure hunting is a gamble, I would warn those lured to its adventure, or those tempted to invest money as backers. And even if gold and silver are found, they have a long history of arousing bitterness and discord among all who covet them.

Treasure diver Tom Gurr pursues his pot of gold in the courts, instead of the sea. Some $500,000 in treasure from a wreck he began working in 1968 awaits division, while Gurr and the State of Florida clash in costly court battles. "Many treasure salvors put $10 or $20 into the sea," says the author, "for every dollar a few take out."

ROBERT MARX

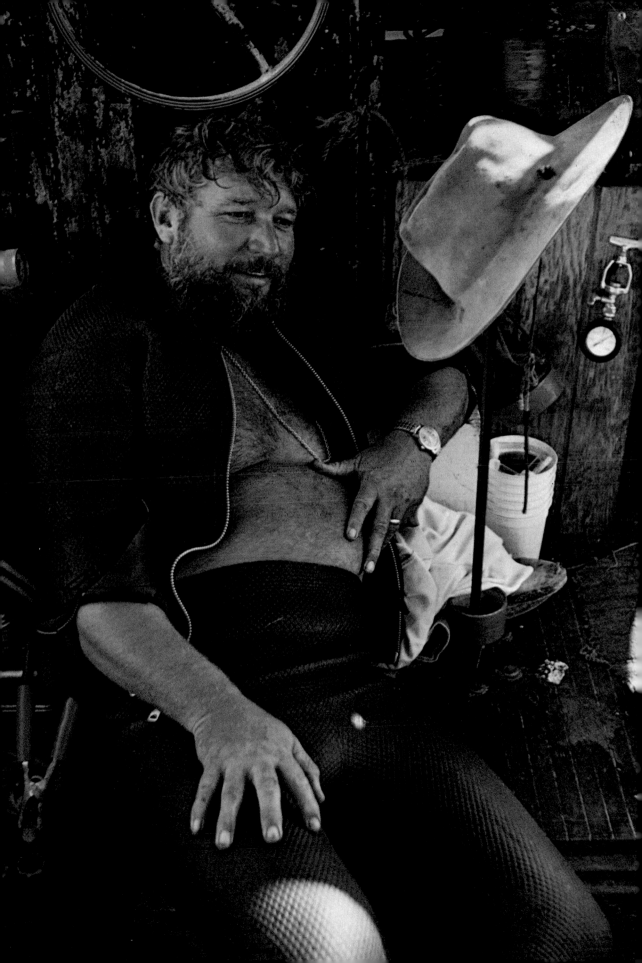

Treasure Hunting: the Amateurs

3

By Robert E. Cahill

GILBERT ARRINGTON is an amateur treasure diver. He has found little gold or silver, but he has recovered brass spikes, old bottles, pottery, chinaware, and other souvenirs. Gil has a mania for searching out such items —but then again, so do I.

As a team, we have looked for sunken riches for about 20 years. For me, it all started when I was 17 years old. My brother, Jim, a Navy frogman, returned home at the

Capsized warship lures divers to Kwajalein Atoll in the Pacific. "Amateurs dive for many reasons," says author Bob Cahill, "ranging from treasure hunting to photographing marine life."

end of the Korean War with his masks, fins, rubber suits, and air tanks—and in need of a diving partner. He taught me how to use scuba, and since then I have soothed my irresistible urge to search for treasure in waters ranging from the cold, dark Atlantic to the coral-carpeted Red Sea.

I was 19 when I met Gil Arrington at Graves Yacht Yard in Marblehead, Massachusetts. He was a boatbuilder at Graves and also the yard's head salvage diver. A carpenter at the yard suggested our first dive when he told us of a British payroll ship that had been wrecked off Crane Beach at nearby Ipswich. "Gold coins dated in the 1700's have been found washed ashore," he told us. "I can show you where."

The following Saturday the three of us drove to the beach. It was early autumn, and the wind and water were cold. But Gil and I wore skin-tight rubber diving suits and were further warmed by the anticipation of finding gold and silver.

"Swim straight out," we were instructed. "There's a reef about a hundred yards offshore—that's where the wreck should be."

Once under the surface, Gil took the lead. He paused to inspect everything and picked up several oysters which he tucked into his suit. We had been under 20 minutes, without a fish or a weed patch to break the monotony, when I gave Gil the thumbs-up sign.

As my head broke surface, a numbing chill flooded my body. The beach was a pencil-line in the distance. Unknowingly, we had been carried 600 yards offshore by an undercurrent.

"Dive and head in," Gil shouted, "the current won't be as strong on the bottom."

I quickly did as he said, but seemed to be making little progress on the sandy floor. My legs began to cramp, and once again I headed up, popped out my mouthpiece and gulped in the cool autumn air. The shoreline seemed no closer than before.

Suffering from cold and nearly sapped of energy, I swam slowly along the surface for about half an hour. Twenty yards ahead Gil had surfaced and was waving me on. Moments later, to my vast relief, one of my fins scraped a sand bar. With Gil huffing and puffing beside me, I swam, walked, and stumbled the final 100 yards to the beach.

"Did you find the payroll ship?" the carpenter shouted as he ran over to the spot where we had sprawled. "Nope," said Gil as he reached into his wet suit, "but you'll have oyster stew for supper."

Gil and I returned to Crane Beach a few years later and, as far as I was concerned, came up empty-handed again. But Gil—who collects anything and everything—retrieved half a dozen enormous clams.

Other edible treasure—lobsters, abalone, and fish—have provided many a meal for amateur divers. But the sea offers considerably more than a dinner, and diving offers more than an ordinary sport.

Anyone who has had the sea close over him knows there is no comparable feeling in the world above the waves. In the depths, life becomes an awesome adventure of search and discovery—with the excitement and the challenge of exploring the unknown.

With the unknown, however, there are also dangers. Some can be mitigated by studying the water conditions and recognizing your own physical capabilities, through training and experience, and by following such precautions as not diving alone.

For many amateurs, clubs provide the opportunity to pursue the sport safely in numbers, to find a diving partner who knows

Overleaf: Coins from the French payroll ship *Le Chameau*, lost in 1725 off Nova Scotia, reward Canadian divers Alex Storm (right) and Dave MacEachern. Wet suits insulate them from the icy Atlantic by retaining sea water warmed by body heat. Boatman Harvey MacLeod shoulders the lifeline that protected the divers against hazardous tides and currents.

BOB BROOKS (OVERLEAF)

61

your strengths and failings below the surface, and to learn from more experienced hands.

Because diving offers something for every interest, whether it's hunting lobsters, exploring wrecks, spearfishing, or photographing marine life, clubs also differ. I know of one group that does nothing but play underwater hockey, while another, the South Pacific Divers Club, has made one ship their specialty—visiting it every weekend for 15 years!

Activities common to most amateur clubs, however, did not interest Gil and me. So we organized a diving group whose sole purpose was treasure hunting. We called ourselves "The Glaucus Underwater Team of Marblehead" for the Greek god of divers.

Our first adventure as a club took place off the Isles of Shoals, a cluster of eight small islands about ten miles off Portsmouth, New Hampshire. It took us more than half a day to sail from Marblehead to the isles in the 45-foot cabin cruiser owned by Captain Ed Saylor of Boston.

Our goal was a frigate—believed to be the Spanish *Concepción*—that foundered off Cedar Island Ledge during a January blizzard in 1813. At the western tip of Smuttynose Island, a nine-foot-high rock monument marks the graves of her Spanish sailors; we anchored about 100 yards offshore.

Gil was first over the side, and I was still struggling into my wet suit when he surfaced and asked for his spear gun. "What

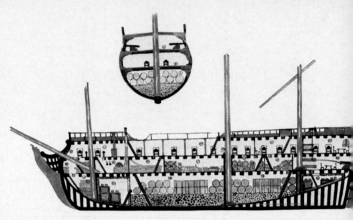

BOB BROOKS; DRAWING BY ALEX STORM

Converted lobster trawler *Marilyn B II* sits high and dry for overhaul on the beach at Louisbourg, Nova Scotia. In 1965 Alex Storm and his partners used the craft to salvage $1,000,000 in coins from *Le Chameau*. Sailing for Quebec with a payroll for French troops, the transport smashed into a jagged reef and sank in shallow water a mile off Cape Breton. Plans of the ship reveal the cannon and stone ballast that marked the wreck—one of the richest ever found by amateurs.

do you want that for?" I asked. "Sharks down there," he replied matter-of-factly, "hundreds of them."

My brother, Jim, grabbed the only other spear gun and joined Gil in the depths. I was contemplating the plunge when Ed Saylor, with a mischievous laugh, pushed me over the side. I adjusted my mask and went down. As I neared the 70-foot depth, I spotted Jim and Gil. Circling them were 50 or 60 sharks, not the man-eater variety, but slinky four-foot sand sharks.

One swam over and nipped my fin. Jim pushed it away with the point of his spear gun, but he poked too hard. Blood gushed from the shark's belly. The others went after their wounded companion, tearing at it with their sharp teeth. That was enough for me and up I went.

I convinced Ed to haul anchor to a spot off Smuttynose near Star Island, where in 1820 three silver bars were found in the shallows. We spent two full days of diving without seeing another sand shark — or any sign of the *Concepción*.

Twice Gil and I returned to the Isles of Shoals. We found no coins, nor did we find a splinter of wood from the Spanish ship. A few years later, however, six divers from Portsmouth found a handful of 18th-century Spanish coins off Star Island at the very spot where we had anchored.

Frustrated by our failures, the Glaucus Underwater Team turned to searching for treasure nearer home, and we found our reward at Graves Island off Manchester, a few miles across Massachusetts Bay from Gil's home in Marblehead. Here, in 30 feet of water, lies the U.S.S. *New Hampshire*, renamed the *Granite State* when she sank. Her keel was laid in 1819, but she waited 40 years for launching. Like the more-famous frigate *Constitution* — "Old Ironsides," built in the 1790's — the ship frame was held together with copper spikes forged in Paul Revere's mill. In 1921 *Granite State*

caught fire in New York and was sold for salvage. While being towed past Massachusetts Bay, the ship caught fire again and sank.

Gil and I found her scattered for almost 100 yards along the rocky bottom near shore. The only salvageable items from her rotting hull were the seven-inch Paul Revere spikes. We collected about 100 of them, all bearing the marking "U. S."

I placed my share in a box in the basement of my home and promptly forgot about them. Two years later, I was offered 50 cents apiece for the spikes. I kept one for a paperweight and sold the others.

Four years later, wandering through a Boston furniture store, I spotted *my* spikes. Converted into candleholders and other knickknacks, most were mounted on polished wood with brass plaques stating that the spikes had been made by Paul Revere and recovered from the U.S.S. *New Hampshire*. Prices for the candleholders ranged from $100 to $150 a pair.

I was heartsick about underestimating the value of my spikes, but I knew they represented a controversy in the diving community regarding the fate of our wrecks — especially the wooden ones. Should they be left intact for other divers to enjoy? Should archeologists evaluate their historical worth before releasing them to salvors?

Some states have already taken steps to regulate the activities of treasure hunters. Unless recovery laws are liberal, however, I am afraid that divers — a secretive group — won't disclose their finds and may even be

Overleaf: Framed by coral-encrusted steel beams, a diver explores the wreck of the *Rhone*. Bound for England in 1867, the British mail carrier broke apart during a hurricane 25 miles from St. Thomas in the Virgin Islands. Near here, the National Park Service maintains two underwater nature trails for snorkelers, and plans similar marine parks for increasing numbers of vacationing divers.

DAVID DOUBILET (OVERLEAF)

65

discouraged from searching for old wrecks. In that case, our maritime past will eventually disappear anyway—lost to shipworms and crashing waves.

Gil had sold most of his Paul Revere spikes too, but he is not the type to be depressed for long. The following summer we were out looking for another wreck. This time the pirate ship *Whidah* enticed us into action. Microfilm records of the crew's piracy trials that I found in the Boston Public Library indicated she carried $3,000,000 in gold, silver, jewels, and ivory tusks plundered from Spanish ships in the Caribbean. On April 26, 1717, her 145-man crew, drunk on wine and apparently further blinded by a thick fog, sailed the *Whidah* into the shallows off Wellfleet, Massachusetts, where breakers crushed her.

The ocean beach near Wellfleet is 12 miles long. Studying the currents, Gil and I judged that the *Whidah* sank off a three-mile stretch flanked by a raging surf and a steep 80-foot sand cliff. We searched the cold, buffeting waters for two days, and as usual—for us and for the majority of amateur treasure hunters—found nothing.

Remembering that hundreds of silver coins had washed ashore in the early 1900's near where the *Whidah* sank, Gil and I called in three Glaucus Team members who owned metal detectors. For three mid-January days we combed the beach. The detectors sounded only once, revealing not the treasure from New England's richest wreck, but a cluster of four flat metal saucers that turned out to be steel sewer caps.

Although there are hundreds of still-to-be-recovered wrecks below the heavily traveled shipping routes from Massachusetts to North Carolina, the East Coast has

Molded vials for perfume (upper) and ink number among hundreds of glass pieces retrieved by bottle hounds from the steamer *Rhone*, 90 feet down in the Caribbean.

no monopoly on sunken ships. And underwater valuables are as diverse as the sites where they are found.

In the western U. S., treasure divers have been heading for the hills, searching stream bottoms for gold nuggets that have washed down since the days of the Forty-Niners. In the Great Lakes, fortunes can be made recovering copper ingots, pig iron, barreled whiskey, and cash in strongboxes. In the rivers and lakes of New York, old boats, muskets, and other artifacts from the Colonial wars have been found. Proving that sunken treasures come in many forms, divers salvaged a 1928 Peerless sedan in good condition from 60 feet down in Lake Pend Oreille, Idaho. Antique car collectors will pay $3,000 for that model.

Next summer I forgot about treasure hunting and ran successfully for the Massachusetts Legislature. The campaign was hardly over when Gil called. "Let's go after the *Whidah* again," he suggested.

"Gil, I'm sick of going on wild goose chases and coming up empty-handed," I replied, "and I'm sick of diving in water that leaves me chilled for days. I'm getting on in age, you know, and the cold bothers me." I was 39 years old and Gil was 50. "Anyway, all the treasure is being found down south."

This was not entirely accurate. One of the great treasure finds in recent times, and certainly one of the greatest by amateurs, took place in the icy waters off Cape Breton, Nova Scotia. There, in 1965, Canadians Alex Storm, Dave MacEachern, and Harvey MacLeod recovered 8,000 silver and 1,000 gold coins worth $1,000,000 from the 1725 wreck of the French payroll ship *Le Chameau*.

The North Atlantic is rich with shipwrecks, and diving clubs, particularly on the New Jersey coast, visit them regularly. But clear southern seas have the big treasures—and warmer waters. "Let's pick a remote island in the Caribbean and find ourselves some real treasure," I suggested.

"I'm all for it," replied Gil. "You choose the place and I'll go."

Of the two million amateur divers in this country, as many as a third will take diving trips outside the continental United States each year. Catering to their wants, a dozen or so travel agencies deal primarily or exclusively with the underwater tourist, packaging "diving vacations" for individuals and groups to all parts of the globe—from the Mediterranean to the South Pacific.

I was determined, however, not only to find my own treasure but also my own treasure island. So for a few months in 1973 I spent most of my spare time looking into promising spots in the Caribbean. One island stood out from all the others—Grand Cayman. It lies with two small sister islands 150 miles south of Cuba. A British possession since 1650, the island is 20 miles long, and from four to eight miles wide. But most important to divers, Grand Cayman is nearly surrounded by a wreck-collecting barrier reef.

Travel brochures make much of Grand Cayman as a former pirate haven: From here buccaneer Henry Morgan sailed to sack the Spaniards at Panama in 1670. Edward Teach, better known as Blackbeard, started his career at Grand Cayman and the English pirate Thomas Anstis lived here for a year after his ship *Morning Star* was wrecked in 1721. At the west end of the island, pirate George Lowther's ship *Delivery* sank.

An article I found in a 1901 Boston newspaper about a local man, Captain James Foster, clinched Grand Cayman as the place for us. Foster reported that "It is easier to find coins off Cayman Island than it is to find sea shells," and he was an expert on the subject. In 1900, while wading in ankle-deep water on the north side of the island, he stumbled upon 58 gold coins, all dated 1753, and sold them for $7,000.

"I found a place called Grand Cayman," I

told Gil on the telephone. "Shipwrecks all over the place and buried treasure, too—used to be a pirate haven—you want to go?"

"Okay," said Gil, "count me in. What will we need?"

"Just clothes. We can rent diving gear there, and it never goes below 60 degrees, so don't bring your long johns," I replied.

"How about boots or sneakers for climbing down the canyon?" he asked.

I paused. "What canyon?"

"Didn't you just say we were going to the Grand Canyon?"

"No," I shouted, "Grand Cayman—an island off Cuba."

"Oh," said Gil. "I didn't think there'd be much treasure in the Grand Canyon."

On December 5 at 8:00 a.m., an unlikely group of treasure hunters met at Boston's Logan International Airport. With me were Mike Aulson and fellow legislators Tony McBride, Bobby Donovan, Jimmie Hurrell, and Bill Ryan. Gil showed up wearing baggy trousers, a blue T-shirt, and red suspenders.

None of them really knew where we were going. An island in the West Indies with treasure on and around it—that's about all they knew. I knew little more, despite my oral embellishments.

Naval archives in Lisbon, Seville, and London provide much information about where ships sank and the cargoes they went down with. But my research, like that of most amateur treasure hunters, was limited by time and money to local libraries, secondary accounts of sea disasters, and romanticized tales of pirate gold.

By 5:00 p.m. that afternoon, we had arrived at our hotel—known in the parlance of the sport as a "diving resort." Hundreds of such hotels in well-established dive areas around the world offer rental equipment, guides, boats, scuba instruction, and occasionally even a darkroom for the underwater photographer.

We settled in for the night and the next day set out to explore the island in a rented car. I was eager to drive to East End where we had been told many wrecks were visible from the shore.

From the sea, one of the openings through the barrier reef that encloses the entire east side of the island is directly off Gun Bay Village. Local tradition says the village name dates from 1788 when ten British ships trying to find the channel wrecked on the reef during a storm out of the northeast. The lead ship, *Cordelia*, under a Commander Popplewell, hit first and fired two cannon shots to warn her sister ships. The other captains misunderstood the signal and all ships sailed in.

We hoped to dive in this area before our week-long treasure hunt ended, but a mile farther up the coast we found an even more intriguing spot. Here, in plain view only 400 yards offshore, sit two large ships. Resting on the reef at a 30-degree list and only partially submerged is the *Ridgefield*, a World War II liberty ship of 9,000 tons. In 1962, traveling from Panama, she smashed into the reef in the dead of night. About 200 yards from the *Ridgefield* rests the 240-foot freighter *Rimandi Midaju*, with her bow section ripped off. She hit the reef in 1946.

The next morning we stopped by Bob Soto's dive shop to rent our gear. "There's no need to use scuba if you're going to dive near shore or in the sounds inside the reef," Bob told us. "All around the island, except on the west side where there aren't many wrecks, the water's no deeper than 10 to 20 feet. And if it's shipwrecks and treasure you're after, the best hunting is inside the reef. Four years ago I found some gold crosses and medallions, and a platinum bar in one foot of water."

Jimmie Hurrell's eyes sparkled—he was getting treasure fever. "Where'd you find this stuff?" he asked.

Bob chuckled. "If there's more down there I'll find it, not you. Don't worry," he

Wet-suited against November seas, members of an amateur diving club prepare to visit the Brazilian freighter *Ayuruoca*. Lost off New Jersey in a 1945 collision with a Norwegian cargo vessel, the 10,500-ton ship rests intact and upright 165 feet down. The rigors of deep diving limit each member to 30 minutes on the bottom in the cold Atlantic. Fueling up for a descent in the 48-degree water, Robert Archambault lunches on a high-calorie sandwich. For added warmth, Michael deCamp pulls on a second wet suit. "If you can dive here," deCamp says, "you can dive anywhere in the world."

71

added, patting Jimmie on the back, "there's plenty for you, too. A lot of divers, mostly amateurs, have found silver dollars and Spanish pieces of eight and pieces of four. Pottery, old tools, and navigational instruments also turn up in the shallows."

Bob showed us two coral-encrusted jugs he had found in just a few feet of water. He thought they were from the 17th century. "Most divers would swim right over these, thinking they were just chunks of coral. Train your eye to the bottom," he instructed Jimmie, who had never been skin-diving before. "Look for straight lines and unusual shapes, and remember that most old wrecks are covered, or partially covered, by coral— which makes them all the harder to find."

We packed our fins, masks, and snorkels into the trunk of the car, thanked Bob, and headed east along the south shore road. As we neared the one road that cuts across the island to the north side, I remembered it was there that Captain Foster had found his 58 gold coins.

So we veered off the shore road and headed across the forested interior. At Rum Point we got out of the car to look at the 30-square-mile area of North Sound. Its shallow waters are protected from wind and sea by a six-mile reef that has only three narrow channels for ships to sail through—a perfect calm-water refuge for pirates.

Near the reef we entered the water. Jimmie was a bit unsure of himself at first, but he quickly became absorbed in the multitude of fish he saw below. Parrot fish, blackeyed squirrelfish, red snappers, golden angelfish, and fat-lipped groupers played in and out of the staghorn coral. Jimmie dived and tried to touch them. Gil, as usual, studied the bottom seriously, and periodically held his breath, dived to the bottom, and picked up some object he thought was interesting.

After about half an hour a heavy rain began to fall. The sun-splashed bottom turned drab and gray so we began swimming the 70 yards back to shore. On the way, I spotted something long and straight on the bottom and dived for it. It was heavy and shaped like a musket, with a curved handle and a knot of coral where a hammer would be. On shore we chipped some of the coral from it, but it was too encrusted to identify.

Early the next day, Bobby Donovan, Gil, and I were on the beach at East End ready with our masks, snorkels, and fins to swim out to the *Ridgefield* and the *Rimandi*. Bobby had been diving only once before, but any jitters he may have had on the way out were washed away by the sights below. Halfway to the reef, in less than ten feet of water, we swam over two large cannon.

We had not moved more than 20 yards from the cannon when we came upon an anchor with a wooden shank half buried in the sand. Gil dived down for a closer look as Bobby and I swam on. In about 15 feet of water, we came to a sand gully that looked like a town dump. The sea had deposited the rotted remains of old hulls, deadeyes, anchor chains, mast rings, and large sheets of metal in the gully. We called Gil over.

It was like watching a little boy on Christmas morning. He made eight or nine dives into the gully, fingering each piece of wreckage as if it were a gold nugget. He found a large collection of cold cream jars and

Souvenirs of East Coast dives decorate the home of Evelyn Dudas, only woman to explore the sunken *Andrea Doria*. A broken cup from the Italian liner lies surrounded by china from an 1886 wreck. Trophies from other ships include cutlery, a figurine, and a "lucky coin" commemorating the opening of the Panama Canal, an eight-inch brass sextant (lying on a family heirloom coverlet), and a porthole with shattered glass that tells of a fire aboard ship. Made into a lamp base, the ship's bell last rang in 1942 when a German submarine sank the freighter *Arundo*—once the *Petersfield*—off the New Jersey coast.

TOMAS SENNETT

slipped four of them into his bathing suit.

As we neared the *Ridgefield* and the *Rimandi*, it became obvious that we could not swim near either the liberty ship or the freighter. Breakers crashed around both wrecks, making them vibrate. But looking up at those ghostly giants from the water was an awesome experience.

Wary of getting too close to the reef, we swam parallel to it from about 50 yards away and soon found a graveyard of wrecks. We spotted two small cannon frosted with coral growth, an anchor, and a piece of metal hull about 80 feet long. Gil spied some old bottles and dived to collect them.

After an hour or so we headed for shore. Gil lagged behind, investigating every little object in the sand. From the time he spat out his snorkel on the beach, to the time we arrived back at our hotel, Gil couldn't stop talking. "I could spend a year out there," he said, scrutinizing his finds.

The bottles were English-made and seemed quite old, but the cold cream jars were worthless as far as I was concerned. Nonetheless, when Gil returned to his room he placed them along with the bottles on his bureau, next to the musket-shaped piece I had found.

His trophies delighted him, and I laughed to myself thinking about all the broken china and corroded silverware that other part-time treasure hunters display so proudly in their living rooms.

On the fourth day of our stay on Grand Cayman rain began to pour and continued into the fifth. But the next morning the North Sound had calmed enough for diving.

Gil and I were eager to search this area, for we had been told that pirates once

Buoyant rope of fish eggs and a 15-pound lobster divert divers in the Atlantic. From Labrador to North Carolina amateurs dine on "crawlers" and other seafood they find living in the wreckage of sunken ships.

MICHAEL DE CAMP

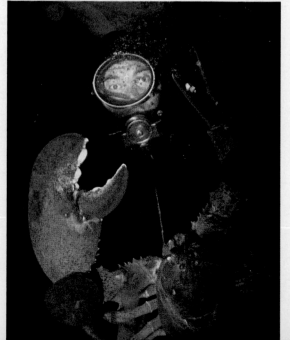

anchored in a cove here. From their anchorage a path led to an inland hideaway. We hoped to find the cove and dive there for coins and artifacts the pirates might have lost in the shallows.

Captain Weston Ebanks had all the scuba and snorkel gear we needed aboard his 40-foot boat as we embarked from the west bank of North Sound. Ebanks kept the boat chugging along parallel to the coast, but we saw no beaches that would indicate a likely landing place for pirates. Then Gil noticed a slight indentation in the growth near a small lagoon.

We anchored and Gil and I were soon over the side in about 15 feet of water, deep enough for a galleon to anchor. Using snorkels, we covered almost every inch of bottom within a 25-yard radius of Ebanks' boat. Except for an occasional waterlogged tree, all we saw was sand, sand, and more sand. I was getting discouraged.

"Let's move farther out into the sound," I suggested. Gil agreed and Ebanks cruised out to about 200 yards offshore. Here, in deeper water, we switched to scuba. We dived and searched around, but found no change in the drab surroundings. After about 20 minutes of uninterrupted monotony, I signaled Gil to surface.

"This is ridiculous," I said. "There are shipwrecks all round this island, and here we are playing in a sunken sandbox."

"Just wait—now wait a minute," said Gil. "I saw a dark shadow out there I want to take a look at." He disappeared under the surface and I followed.

Gil's dark shadow turned out to be a large turtle, swimming leisurely along the bottom. Its shell was about as big around as an automobile tire and part of its left hind flipper was missing. We tried to remain still and out of sight, but it saw us and hurried on. I was surprised that an injured turtle could swim so fast.

It headed for a pile of rocks and wedged itself into a crevice. Gil swam over and touched its shell but the turtle didn't move. I thought about going for a ride on its back, but Gil's thoughts had turned away from the turtle to the rocks it was hiding among.

I too forgot the turtle when I realized the rocks were ballast stones off an old ship! Gil started pulling some of them from the pile, and under one he found broken glass—the remains of a rum bottle. I began helping him tumble them off the pile.

Running low on air, we surfaced and swam to the boat. Gil borrowed a coil of line and a buoy from Ebanks and swam back to the rock pile to mark the spot.

"What did you do that for?" I asked him. "We're leaving tomorrow and won't have time to get back here."

"I'm not going home," said Gil, lifting himself back into the boat.

I was stunned for a moment, then laughed. "You're not serious are you?"

"I have nothing tying me to Marblehead," Gil replied. He was smiling now, a smile of revelation.

"You can't live on rum bottles and cold cream jars," I pleaded.

"I bought half an acre of land yesterday," he said, "and I can pitch a tent or build a shack there. I talked with a fellow who's willing to help me build glass-bottomed boats. Think of all the tourists who can study life on the reefs and look at shipwrecks without getting their feet wet."

The next day Gil drove me to the airport, and the last time I saw him he was standing at the terminal gate in his baggy trousers, blue T-shirt, and red suspenders.

As my plane took off, I thought about Gil renting his glass-bottomed boats and wondered how many people, excited and challenged by the world beneath the waves, would themselves someday be diving in search of sunken riches—and I knew that I would always be among them.

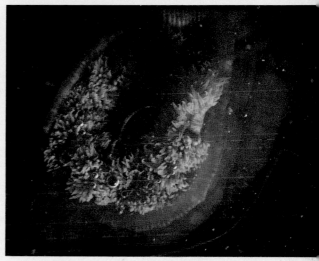

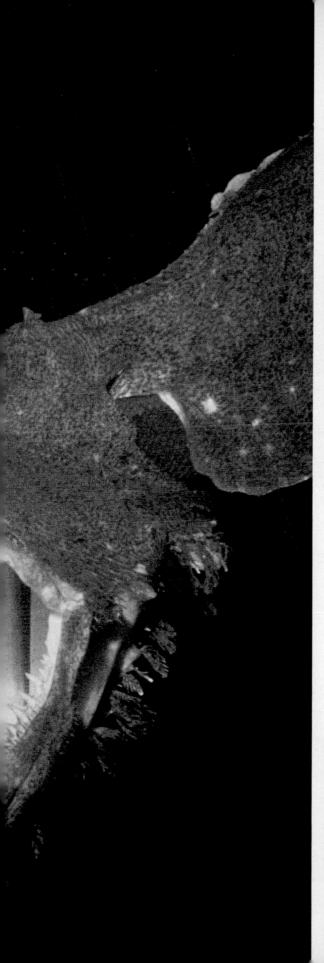

Sunken vessels provide niches for marine life off the East Coast. Diver-photographer Michael deCamp cautiously holds a goosefish, also called an allmouth or monkfish; with rows of jagged teeth, it can inflict a deep wound. Like the giant pollock (below), it preys on fish that find refuge in shipwrecks. Eel-like ocean pouts (top) nestle between the steel plates of an oil tanker. Like the sedentary sea anemone (above), they feed on small crustaceans. Protected by stinging tentacles, this anemone grows on the hull of a freighter.

In eerie light 15 miles off Point Pleasant, New Jersey, Edward Rush—one of the first divers to reach the Norwegian tanker *Stolt Dagali*—glides along her deck and past her monogrammed smokestack. On November 26, 1964, the Israeli luxury liner *Shalom* cut through the *Stolt*. The tanker's bow remained afloat, but her stern plunged to 130 feet with 19 crew members. One year after the collision, the *Stolt* almost claimed more lives: While exploring the wreck, two amateurs ran out of air and made emergency ascents. Above, one rides a litter from the deck of their dive boat to a hovering U. S. Coast Guard helicopter that answered their SOS. Annually the Coast Guard responds to about 100 such calls from divers in distress.

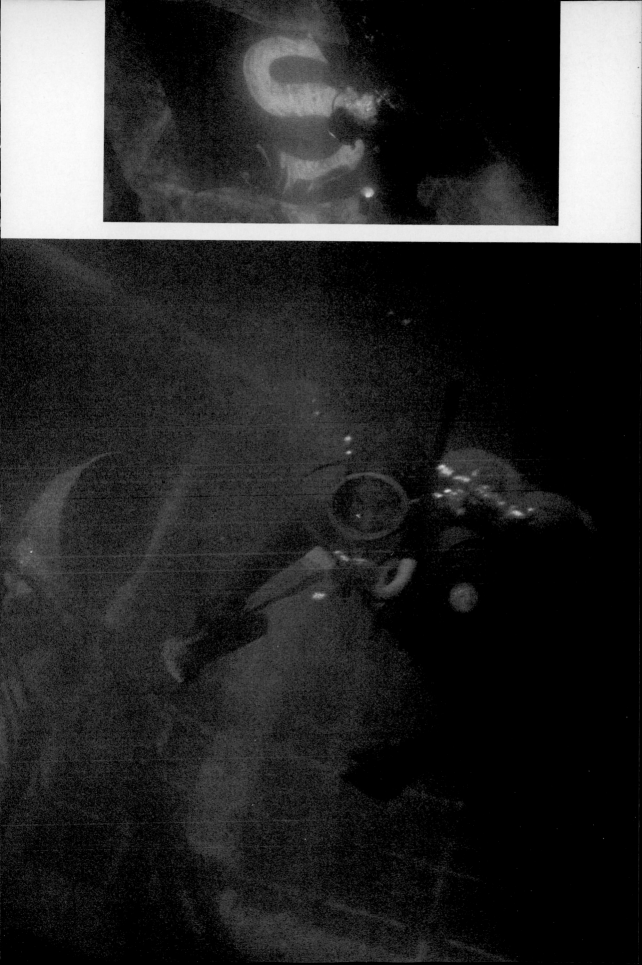

ROBERT MARX

Treasure Hunting: the Professionals

4

By William Graves

THROUGH a raging sandstorm 40 feet under the sea I grope for the bottom and dig in with both hands, fingers arched like grappling hooks. Blinded by torrents of sand against my face mask, I recall Dean Curry's advice before the dive: "Don't waste your time trying to see down there, because you can't. Once you touch bottom use the current as your compass—swim directly into it and it'll lead you to clear water."

Cautiously I haul myself deeper into the

Reason to smile: A 7½-pound disk of solid gold —part of a multimillion-dollar find from a 1715 Spanish wreck—provokes delight in treasure divers Fay Feilds and Mel and Dolores Fisher.

storm, hand over hand across the ocean floor. Something large and solid suddenly brushes against my shoulder and I reach out for it, but there is nothing, only the swirling sand.

Gradually the silence of the undersea is broken by a rhythmic thunder that seems to come from somewhere ahead. The thunder grows louder, reaching a point directly above me, and all at once the curtain of sand dissolves.

I am on the slope of a large crater measuring perhaps 30 yards across at the rim. Below me are two divers, one of them Dean Curry. Through clear water overhead I glimpse the thunder's source: the "mailbox," salvor Mel Fisher's ingenious brainchild. The dark hull of a ship is moored securely by a network of anchor lines, its churning screws housed in two enormous steel tubes mounted to deflect the wash downward in a powerful jet of clear water that drills slowly into the ocean floor.

Descending along the crater wall, I join Dean near the bottom. He waves a welcome and continues to search the surrounding sand as it dissolves beneath us and is swept out over the crater rim. I watch, fascinated, as scores of brightly patterned shells and lumps of coral emerge from the sand and slowly take shape like a vast undersea garden in bloom.

At the bottom of the crater Dean suddenly taps my shoulder, pointing to a small black stain on the beige surface of the sand. He reaches into the center and removes a small, thin object. Rubbing it between the fingers of his glove, he holds it out for me to see—a badly corroded but recognizable Spanish piece of eight.

Since that moment beneath the sea off the Florida coast, now many months ago, I have suffered from an ailment common among treasure hunters—the disease known as *febris auris,* or more commonly, "gold fever." So far I have found no gold myself, though I have seen a great deal of it recovered by others.

Febris auris is as old as man himself, and history records countless epidemics of it: California in 1849; South Africa in 1886; the Klondike in 1897, to name just a few. Perhaps the most violent epidemic of all, and the reason for my dive with Dean, swept the continent of Europe between the 16th and 19th centuries and turned parts of the New World into a giant lodestone that beckons men to this day.

Nowhere is gold fever more rampant and incurable than among professional salvors, those whose lives and fortunes are spent— and occasionally lost—endlessly searching for sunken treasure.

One man who contracted the fever early in life is Melvin B. Fisher, the president of Treasure Salvors, Incorporated, and Dean Curry's boss. A veteran diver and explorer of more than 100 wrecks, Mel Fisher has spent nearly half his 52 years in the search for sunken treasure, with results ranging from bonanza to near-bankruptcy.

Today Mel directs the world's largest undersea treasure salvage operation, with three large ships, 22 workboats, and a staff of 40—divers, technicians, guides, mechanics, and an office staff.

My dive with Dean resulted from Mel's invitation to join one of his crews for a day in the Marquesas Keys, some 70 miles to the southwest of Florida's tip. Here Dean and his fellow divers have spent several years salvaging a wreck that Mel believes to be the legendary Spanish treasure galleon, *Nuestra Señora de Atocha,* lost with a fortune in gold and silver during a hurricane

(Continued on page 90)

Overleaf: All in the family, Mel Fisher's co-workers include son Kane, here probing sands near the Marquesas Keys, Florida, for the *Nuestra Señora de Atocha,* a rich treasure galleon Fisher believes sank here in 1622.

N.G.S. PHOTOGRAPHER BATES LITTLEHALES (OVERLEAF)

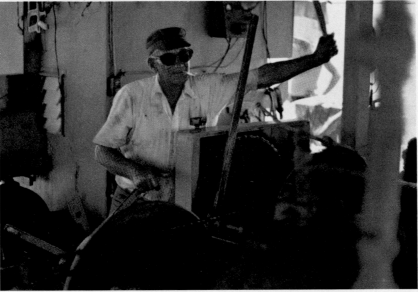

Bringing home the daily hoard, Fisher's divers load tarnished silver bars and a coral-encrusted barrel hoop onto a replica of a Spanish galleon—museum, office, and strong room for Fisher in Key West. Head of the world's largest undersea treasure salvage operation, Fisher once enlisted the help of Capt. Art Hempstead (above) and his airlift (below), a sort of undersea vacuum cleaner that clears sand and debris from bedrock. But even the airlift proved too puny to move the Keys' deep sands, and the team began using deflected propeller wash.

Blizzard of bubbles and sand envelops a diver and
his red-tipped snorkel as he works the bottom
beneath the "mailbox," a Fisher invention named
for its resemblance to a letter chute. First designed
to bring clear water to murky sites, the mailbox
diverts prop wash to the sea floor, where it digs
through sand and mud blanketing treasure. In little
more than half an hour it can cut through sand
20 feet deep. Divers hover nearby, darting in to
snatch artifacts as they turn up.

One of the richest prizes in the Atlantic, the
Atocha sank off Marquesas Keys, Fisher believes,
where shifting sands gradually bury wrecks even
today (below right). Old Spanish records found by a
colleague of Fisher's tell of the *Atocha*'s going down
off "the last key of Matecumbe"; but the Spanish
often called the entire chain of Florida Keys the
Matecumbes, not just the two we know today.
Despite Fisher's theory, other treasure divers still
believe the *Atocha* lies 100 miles to the east, off
Lower Matecumbe Key (below).

GEOGRAPHIC ART

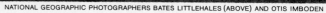

NATIONAL GEOGRAPHIC PHOTOGRAPHERS BATES LITTLEHALES (ABOVE) AND OTIS IMBODEN

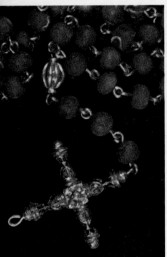

Businesslike as bank tellers despite the flash of gold, Dolores Fisher and State Underwater Archeologist Wilburn Cockrell catalog part of the Marquesas treasure, destined for safekeeping in Tallahassee. Florida claims every wreck in her waters and retains all items recovered until divers have finished their work; salvors then receive 75 percent of the find. Artist and ship's cook, Jo Arden Fisher (opposite) sketches a rosary from the Marquesas site. Other finds include a hand-stamped gold coin minted in Seville and a crushed, but finely chased, gold bowl of Oriental design.

in 1622 on the return voyage from Havana to Spain.

Up to the time of my dive with Dean, Mel's crew had brought up numerous 17th-century Spanish artifacts, four gold bars, three 65-pound silver ones, and more than 7,000 silver coins and pieces of eight such as the one I watched Dean recover.

With only the single coin, Dean and I surfaced soon afterward. Once aboard the salvage ship *North Wind,* I asked him about the black spot on the sand below.

"It's a stain caused by silver sulphide from the buried piece of eight," he said. "Over a long period the metal oxidizes, or tarnishes, and stains the sand around it. That's why silver coins are often easier to spot than gold ones, because gold doesn't tarnish."

With the ship's propellers stilled, the water became crystal clear again and I noticed a sizable black shape hovering near the crater below.

"That's the ship's mascot," Dean explained. "He's a four-foot barracuda, and we call him Ralph. He follows us everywhere, feeding off small fish attracted by the disturbance below. Mind you, Ralph thinks we're working for *him,* and he shows his gratitude by rubbing against the divers and poking his nose into their work. Too bad you didn't meet him."

I recalled my encounter with something large and solid during the sandstorm below and had a sudden suspicion I'd already made Ralph's acquaintance.

Back at Treasure Salvors' headquarters in Key West I learned some of the major problems of the trade from Mel and his historical consultant, Dr. Eugene Lyon.

"The treasure business has changed enormously in recent years," Mel told me. "I first began salvage diving as a hobby around 1954 while I was operating a dive shop in California. I had a 65-foot dive boat, some scuba tanks, and a lot of persistence. I did my own research, talked with the local fishermen and old-timers along a stretch of coast, picked a likely wreck, and began diving."

I asked if it was true, as I'd heard, that Mel's wife, Dolores, paid for groceries in the early days with pieces of eight that Mel had salvaged. He shook his head.

"It never quite came to that," he answered, smiling, "but we did pay for a new roof on the house that way and I once bought a car for Dolores with some gold doubloons.

"Nowadays," he added, "treasure salvage is a huge and complex business. In my case, it involves running a fleet of ships, keeping accounts for state and federal taxes, operating the latest electronic detection and recovery gear, making sure there's a state observer aboard each of my salvage craft, and trying to keep everything going." He nodded at Gene Lyon. "And there, probably, sits the key to it all.

"If it weren't for Gene we'd never have found the Marquesas site. For years historians believed that the *Atocha* went down off what are known today as the Matecumbe Keys, halfway along the chain between Key West and Miami. Early accounts of the wreck, you see, stated that the ship went down, 'on the west side of the last key of Matecumbe. . . .'

"But Gene spent more than a year digging through the Spanish archives at Seville, where they keep most of the original documents from the days of the New World fleet system. He discovered something that others had overlooked: The Spanish often referred to the entire *chain* of Florida keys as the Matecumbes.

"If you look at it that way it alters the whole meaning of the *Atocha* account and shifts the wreck site more than 100 miles to the west. I know that some salvors aren't convinced we've found the *Atocha* site, but I am, and we'll keep on working the site until we hit the main treasure."

From Gene I learned that laws of the past concerning gold and silver were just as strict as those of the present, and just as regularly broken.

"No one knows exactly how much gold and silver came out of the New World aboard the Spanish treasure fleets," he told me, "and even the Spaniards themselves never knew. Some estimates run as high as 14 billion pesos' worth of registered treasure over the space of more than three centuries, and maybe that much again in contraband, but that's only guesswork.

"Of course," he added, "the Spaniards kept careful records, and they stamped really valuable items such as gold bars with identifying marks and serial numbers. The authorities searched every vessel at both ends of the voyage and sometimes even under way, to discourage the captains and crews from smuggling." He shook his head. "But a lot of gold slipped past them, and almost every galleon was carrying more treasure than her record showed."

Ironically, a metal more precious today than gold existed in the New World, yet for years the 18th-century Spaniards would have none of it. The substance was platinum, and grains of it were often found mixed with gold in mines in Colombia. Two gold mines were even abandoned because of the high platinum content of the ore. Dropped into molten gold, the heat-resistant grains would become irretrievably stuck together, ruining the gold. Ordered to dispose of the platinum at the mines, the Spaniards set slaves to painstakingly separating the two metals grain by grain with knife blades. Then the King's officers, accompanied by witnesses, took the platinum to the Bogota River and threw it in.

I heard the story from Art McKee, the veteran salvor affectionately known as the grandfather of treasure hunters. Despite a head of silvery white hair, Art at 62 has more energy and ambition than many a man 20 years his junior. When I paid a visit to

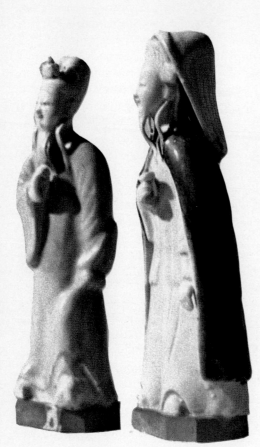

his popular Museum of Sunken Treasure on Florida's Plantation Key I found him deep in plans for still more treasure-hunting expeditions, in the Caribbean and off Spain. Art kindly put his work aside for an hour or so to give me a tour of the museum.

We began with an array of hard hat diving helmets and progressed through Art's collection of treasure and artifacts recovered from some 300 sunken wrecks he has explored over nearly half a century. There were cannonballs, arquebuses, sextants and octants, tableware, china, lockets, and hundreds of ship fittings. As we walked slowly among the showcases, Art referred by name to the ships that had produced the collection—*San Pedro, San José, Capitana Rui, Looe*—not merely as vessels, but rather as cherished shipmates of long-ago voyages and adventures.

We came at last to the gold room, a large walk-in safe lined with displays of glittering crucifixes, jewelry, doubloons, bars, and disks that seemed to fill the room with soft fire. I remarked that it was quite enough to give anyone a severe case of gold fever.

"Nobody's immune," Art agreed, nodding, "least of all the old-time Spaniards. They went to incredible lengths in the search for gold and they didn't let human life—especially somebody else's—stand in their way. During three centuries, tens of millions of New World Indians died, mostly from European diseases and famine, but many in battle. A few million were simply worked to death in the gold trade. Indians mined the gold, loaded it aboard galleons, and when the galleons went down, the Indians—usually Caribs, who were excellent

True to their name, two of the Eight Immortals—saintlike humans from Chinese theology—survived a 1781 sinking off South Africa. Salvors Reg and Bill Dodds recovered all eight from the *Middleburg*, a Dutch East Indiaman carrying Ch'ing Dynasty porcelain.

divers—were there to salvage the cargoes under Spanish direction. By 1550 the Caribs were nearly extinct, and African slaves, also accomplished divers, took their place.

"A lot of people," Art continued, "think that when a galleon went down, the Spaniards simply wrote it off, but they didn't. They were geniuses at the art of salvage, and if a wreck lay in less than 60 or 70 feet of water they sent the slaves down to get it. A good many of the treasures we salvage today by modern techniques are simply leftovers from those early recoveries."

Now and then in the salvage world success comes disguised in the form of a spectacular failure. None in recent years has been more spectacular than that of Don Rodocker and Chris DeLucchi. Don and Chris were after one of modern history's most famous and elusive treasures: the ship's bank aboard the ill-fated Italian liner *Andrea Doria*. Although the two divers recovered virtually nothing of value, the attempt made salvage history and guaranteed them both a fortune in years to come.

I met Don and Chris in August 1973, in the quiet coastal town of Fairhaven, Massachusetts. They had just returned to port after their attempt on the *Andrea Doria*, but our meeting was delayed several hours: Both men were undergoing decompression inside *Mother*, an undersea habitat they had designed and built themselves.

Don's and Chris's attempt to salvage the *Andrea Doria* had already attracted national attention. The ship itself was famous, victim of a nighttime collision in 1956 with the Swedish liner *Stockholm* some 50 miles south of Nantucket Island. *Stockholm* escaped with major damage, but the crash took 50 lives and sent *Andrea Doria* to the bottom of the Atlantic.

During the next 17 years professional and amateur treasure hunters alike sought either to raise the liner or to reach the ship's bank, reported to contain more than two million dollars in jewelry and cash. Prior to Don's and Chris's attempt several prospective salvors had lost their lives on the wreck site or in practice dives elsewhere.

With early scuba, only the most experienced and skillful diver could cope with the currents, the cold, the sharks, and the depths that had put *Andrea Doria* almost beyond human reach. Don and Chris were both crack U. S. Navy divers and they not only reached *Andrea Doria*, they also lived and worked aboard her for eight days.

Their secret was *Mother*, a device that closely resembles an oversize steam boiler still enclosed in its shipping crate. Nearly 13 feet long and weighing slightly over ten tons, the undersea capsule can support three working divers for two weeks at depths as great as 600 feet. In an era when research alone for undersea vehicles and habitats generally runs to millions of dollars, *Mother*'s total cost of $65,000, plus $85,000 in operating costs for the dive, opens a vast new field to professional treasure salvors—the technique known as saturation diving.

From the pier at Fairhaven I watched as *Mother*, having surfaced from the wreck site at sea by means of her own ballast tanks, was towed into port by a converted fishing trawler and winched out of the water. One of the handling crew explained to me that the capsule had been attached to *Andrea Doria*'s hull at a depth of 160 feet; and that U. S. Navy tables called for 68 hours of decompression, but new charts developed especially for this dive cut the time to less than half that.

Finally *Mother*'s hatch swung open and Don and Chris emerged with the expedition's chief photographer, Bob Hollis. At an impromptu press conference the two men, only 27 and 22 respectively, sketched the history of the dive and explained their satisfaction with the results.

Mother had been designed and built over

"One of the most treacherous, God-forsaken spots in any ocean on this earth," says salvor Roland Morris of Gilstone Rock in the Scilly Isles. Here, amid half-submerged crags and saw-toothed reefs, Morris found the battered *Association*, a British man-of-war sunk in 1707. He recovered thousands of coins, as well as three elaborately worked French cannon (left). Positive identification of the wreck came from a silver plate engraved with the crest of the Admiral, Sir Cloudesley Shovell.

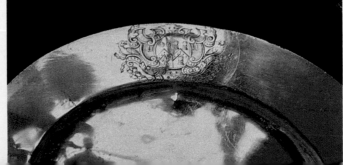

the previous year in San Diego, California, and then trucked across the country on a flatbed trailer. Towed to the wreck site, she had been lowered after a series of difficulties and the two divers had set about cutting through the hull with an underwater torch. For most of the time they spent below, the men had breathed a mixture of 5.4 percent oxygen and 94.6 percent helium through "umbilicals"—long hoses that extended from the habitat and gave *Mother* her distinctive name.

Even after 17 years the sunken ship proved surprisingly free of marine growth. As Chris DeLucchi remarked, "When we came alongside in the habitat we could see the white stripe of the Italian-American Line clearly on the 700-foot hull."

Life on *Andrea Doria* proved difficult and frustrating. Plagued by ear and skin infections, by the cold, and by occasional trouble with their mooring system, the two divers were constantly delayed. Finally, on the morning of the eighth day, they managed to cut through the hull doors into the foyer of the ship's bank—and instantly realized that the expedition was over.

"There was no way through it," Don said of the compartment. "After 17 years of submersion and rot, the wood paneling of the foyer bulkheads had collapsed and jammed the passageway—it was solid debris from end to end. The paneled ceiling had given way and was barely hanging by electric cords to the light fixtures. Just the pressure from our air bubbles at that depth might have caused it to collapse. It would have been suicide for any diver to risk it."

Returning to the habitat, the two men talked it over. Although they were only a matter of yards from a possible fortune, neither one would consider the use of explosives. "If we had used them to clear the passage," Chris explained, "the ship's oil tanks might have ruptured, and we've both seen what that can do in California. Our

infections were getting worse and time was running out on us: It was late August, the season for hurricanes in New England."

Reluctantly the two signaled the surface for support divers to come down and release the cables that secured *Mother* to the ship, and the habitat made its smooth ascent. For all the disappointment, *Mother*'s maiden voyage had been a success, for it had demonstrated that saturation diving—the technique of prolonged existence in an undersea habitat—could be economically applied to treasure salvage.

Any lingering doubts about their invention were dispelled when Chris and Don returned to San Diego. Within a week or two their recently formed company, Saturation Systems, Incorporated, was besieged with requests from insurance and salvage firms to put *Mother* to work retrieving millions of dollars' worth of sunken cargoes.

By contrast to *Mother*'s sophisticated design, the 42-foot workboat *Brigadier* is downright primitive. Broadbeamed, flat-bottomed, and equipped with a venerable air compressor, she ambles tirelessly among the great coral reefs off Bermuda in search of their historic victims. Yet the appearance is deceptive, for *Brigadier* and her master, a 49-year-old Bermudian named Teddy Tucker, are among the most famous and successful teams in the entire field of professional salvage. One reason is the small item known throughout the trade simply as "The Emerald Cross."

No one knows precisely how the cross reached Bermuda, or even the name of the ship that carried it. Teddy believes that the crucifix must have belonged to an official of the Catholic Church, most likely a bishop or higher, and that its distinguished owner was already dead by the time his ship came to grief.

I first saw the cross with Teddy in the small museum at The Flatts Village, east of

HARRY TRASK, BOSTON TRAVELER (BELOW) AND JACK McKENNEY, OCEANIC FILMS

Trailing a shroud of foam, the *Andrea Doria* (above) slides toward
the bottom with an estimated 60 million dollars in salvageable
treasure and scrap. She sank in 1956 after colliding off Nantucket
with the liner *Stockholm*. Many divers have visited the 240-foot-deep
wreck—but briefly, for massive pressures at that depth force poten-
tially fatal amounts of nitrogen into the blood and necessitate long
periods of decompression. In 1973 American divers Don Rodocker
and Chris DeLucchi (top) solved the problem by staying below,
living between dives in a pressurized submersible chamber. Called
Mother, the chamber (left) holds diving gear and emergency life-
support systems. Hoses from equipment aboard the converted
trawler *Narragansett* (top left) supplied *Mother* with electricity,
heat, and breathing gases; the divers survived *Andrea Doria*'s deep,
cold grave for eight days while they cut and entered her hull.

JACK MCKENNEY, OCEANIC FILMS

Hamilton, Bermuda's capital. Teddy recovered the cross in 1955 and sold it along with his other treasures to the Bermuda government for $100,000. Earlier, he had refused an offer of $75,000 for the cross alone.

Measuring about three inches long and two across, the crucifix glows with the richness of gold and the luster of seven perfectly matched emeralds totaling 35 carats in weight. To Teddy, however, the fascinating thing about the cross is not what it contains but what is missing from it. Pointing to the pair of beautifully engraved arms, he indicated a small ornament of gold suspended from each like a tiny nail or peg. My first thought was that the ornaments had once held additional stones or perhaps pearls, but Teddy shook his head.

"They represent the nails in Christ's cross," he said, "and the traditional number was three. Now look at the base of the crucifix and you'll see a small hole where the third nail should be, but isn't. The chances are a thousand to one against that nail's having fallen off. The Spaniards did excellent work, the cross isn't damaged, and the two other nails are still in place. The third nail must have been *taken* off, a practice followed in former times to signify the owner's death. In all likelihood the man who wore that cross was on his way back to Spain for burial when a storm caught his ship and buried him at sea instead."

Teddy's bishop is hardly alone, for Bermuda lies in the midst of a vast marine

Carving his own doorway, Don Rodocker slices into *Andrea Doria*, hoping to find the purser's safe, ship's bank, and jewelry shop. But inside he found it "all caved in, a death trap"; tons of wooden paneling dangled from light cords. He and DeLucchi admitted defeat, retrieving only a corroded serving dish cover (left) and other items of little worth. *Mother* proved invaluable, however, and marine insurers now deluge the pair with salvage contracts.

graveyard. Throughout centuries of sail the former British colony served both as beacon and burying ground for countless ships that approached Bermuda to pick up the prevailing westerly winds for the return to Europe. Not all the victims were galleons or even Spanish, as I learned during a day's diving with Teddy and his partner, Bob Canton. Setting off in *Brigadier* with a third member of the crew, Teddy's seven-year-old Schnauzer, Mitzi, we explored several wreck sites from among more than a hundred that Teddy and Bob have successfully mined over the past 25 years.

Through somber galleries of coral, with a glittering escort of tropical fish, we dived first to the *San Antonio*, a Spanish treasure galleon driven to her death on an offshore reef by a hurricane in 1621.

Never one for mere sight-seeing, Teddy combined our inspection dive with a little work—and once again I observed the difference between amateur and professional treasure hunters.

Though the site had been mined, and seemed to me utterly barren, Teddy plucked from the sand a tiny sphere of iridescent blue and held it to my face mask. It had been molded with a neat hole through the center, fashioning a glass bead for a necklace—a trading item for the Indians. Three more beads followed the first before Teddy gestured upward and we surfaced.

In all, we visited four wrecks that day, including another galleon lost in the 1650's, then a ship known to Teddy simply as "Tankard" after a pewter mug he found aboard her, and finally a more recent casualty, the British brig *Caesar*. It was lost in 1818 with a cargo of massive grindstones and bricks for a church in Baltimore.

With the born salvor's instinct for a market Teddy retrieved several hundred of *Caesar's* grindstones and sold them to Bermuda resort hotels to be laid out in picturesque terraces.

Around the world, few wreck sites have a more spectacular setting than South Africa's lonely west coast. Fewer still have a history as deeply etched by violence and tragedy at sea. Almost since the dawn of sail the great ocean portal has confronted men with the challenge implicit in its name, Cape of Good Hope. More often than not, hope has been the last recourse of sailors beset by one of the fearful Cape storms, with monstrous seas rolling up from Antarctica. Among South Africans the western approach to Cape waters is justly known as Skeleton Coast, for some 1,200 wrecks are said to lie there. Along that desolate coast two brothers named Reg and Bill Dodds have recently made salvage history.

Thirty-one and thirty-three respectively, the Dodds brothers are seasoned professionals with roughly a quarter of a million dollars' worth of salvaged treasure to their credit, yet like Teddy Tucker they work with the simplest of gear. Their richest strike to date was made with secondhand diving equipment and a retired abalone boat. One item rarely used by American salvors is vital to their operation—a submersible iron cage for protection against the Cape's dreaded blue-point and great white sharks.

Happily for me the cage was unnecessary during my visit to Skeleton Coast with Reg and Bill. The area most heavily patrolled by sharks was besieged by a waning autumn storm, and even the Dodds brothers thought better of diving.

On such a day not long before, the brothers had anchored their boat some 75 yards offshore and Bill had dived on the 18th-century Dutch East Indiaman *Meresteijn* while Reg tended the boat. For an air supply Bill used the "hooker," a long hose connecting his mouthpiece to the boat's compressor.

"It was a rough day," Reg recalled, "with 15-foot waves exploding on the rocks ashore. Bill reached the bottom about 20 yards inshore of me, when suddenly the boat's anchor began to drag. There was no way to stop it, and I knew that within minutes we'd lose the boat, the gear, and everything on the rocks. I couldn't signal Bill and there was only one thing to do—I cut Bill's hose, pulled up the anchor, and took off for open water."

With a single breath of air Bill somehow managed to swim 50 yards to shore, waited his chance, and scrambled to safety between waves. I asked if he'd been angry with Reg for deserting him, and he merely grinned.

"Letting the boat break up on the rocks probably wouldn't have saved me," he said. "I think Reg made the only possible decision. If I'd been in his place I'd have done the same thing."

Abandoning the *Meresteijn* site, we turned to the comparatively sheltered waters of Saldanha Bay, some 70 miles northwest of Cape Town, where on a memorable day in 1969 Reg and Bill brought up a small fortune.

"We were diving on *Middleburg*," Bill explained as we crossed the bay in their 16-foot outboard. "We'd discovered the wreck two years before but let her sit while we worked another site. In those days, you see, we were strictly after coins, and *Middleburg* was a Dutch East Indiaman bound for home."

I failed to see the point, and Bill explained further about East Indiamen. "They were usually English or Dutch," he said, "sailing for the two great companies, British East India and Dutch East India. During the 17th and 18th centuries they practically monopolized the trade between Europe and the East, bringing back fortunes in spices, silks, and dyes. The trade was so incredibly rich it was said that either company could make a profit if only one in every *five* of its ships got through.

"But to get the silks and spices," Bill continued, "of course they needed money, and they took it out by the chestful. As a rule, the eastward-bound ships carried coins and bullion, while the ones coming west carried cargo."

Under the circumstances Bill and Reg had their doubts about *Middleburg,* for she was not only westbound but also had been deliberately destroyed by fire.

"The year was 1781," Reg continued, "and Holland and England were at war with each other over your American Revolution. The *Middleburg* was surprised by a British naval squadron at Saldanha Bay, and her crew set her afire and scuttled her rather than let the enemy have her."

But for that long-ago decision *Middleburg* today would be worthless as a treasure wreck, for she was carrying a cargo of china. Had she put up resistance the British men-of-war would have blown her apart and left nothing of her cargo but colorful fragments. Instead she drifted gently aground in a protected cove where Reg and Bill finally dropped anchor.

"I'd been down for several hours with nothing to show for it," Reg said of the day *Middleburg* relinquished her treasure. "I'd seen a lot of porcelain chips, which were certainly nothing new, and I'd about had it. Then I saw something larger poking out of the sand, so I pulled it loose and stared at it." He paused at the recollection. "And you know, the thing just stared back at me."

In his hand Reg held an exquisite work of art, a small porcelain figurine of the Ch'ing Dynasty, successor to the famous Ming period in Chinese culture. Later study by experts identified the statue as one of the Eight Immortals from Chinese theology. Even before he knew the full extent of his find, Reg made a second one in the shape of another Immortal lying beside the first. With a statue in each hand he shot to the surface and yelled for Bill.

"What did you say?" I asked.

"What would *you* say?" he replied. "I yelled, 'We're rich, we're rich!' "

Subsequent dives brought up other Im-

mortals until Reg and Bill had nine, one of them a duplicate. Taken to Cape Town, they were eventually sold for nearly $10,000 to a South African art collector.

"If we'd sent the statues to London or New York for auction as we did some of the later finds," Bill said, "they'd have sold for a great deal more. But then, when we first began in this business we were so green that we sold 17th-century Dutch coins to local jewelers to be melted down at the standard price of 75 cents an ounce for scrap silver."

Later during my visit Reg and I went down for a look at *Middleburg,* or all that remained of her. Even through our heavy neoprene wet suits we could feel the chill from the icy water. Our diving companions, a number of jackass penguins native to the Cape area, made it seem even colder.

Despite a murky bottom from dredging operations nearby, I saw several large pieces of Ch'ing porcelain and one of *Middleburg*'s badly corroded cannon. It occurred to me again that if it had been fired against the British on that day in 1781, the fire doubtless would have been returned and the Eight Immortals might not have lived up to their name. As it was, the world was indebted to that long-ago Dutch crew who had chosen a heroic but gentler death for their ship.

I left Cape Town soon afterward, for Reg and Bill were busy with a film—"Pieces of Eight"—about their adventures. As we said goodbye Bill revealed that singular blend of hope and eternal optimism common to all great salvors.

"There's a wreck up the coast," he confided, "that puts all the others to shame. She was a British East Indiaman, the *Grosvenor,* and what a cargo—solid gold throne, chests full of jewels, boxes of bullion, the whole lot. People have tried for a hundred years to find her without any luck, but Reg and I will do it, true as God. We'll send you a cable in America one day...."

I'm confident they will.

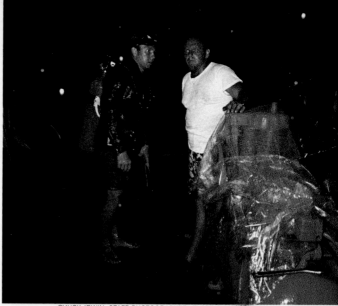

Fearsome symbol of power and godliness in the Maya world, a stone jaguar begins its ascent to daylight after centuries in the dark, water-filled Sacred Cenote in Chichén Itzá, Yucatan. Innovative leader Norman Scott (above, at right) finances his global expeditions with grants from U. S. industry and the sale of movie, book, and magazine rights. He used chemicals (below) to precipitate the heavy silt in the Cenote. From the cleared water, divers brought up more than 6,000 artifacts, including the bones of sacrificial victims, pieces of copper, jade, gold, incense, and two wooden stools, the first Maya furniture ever recovered.

Archeology in Old World Waters

5

BY PETER THROCKMORTON

WAVERING SHADOWS played across the Turkish sponge diver's features, already distorted by the glass faceplate in his helmet. But I could sense his pleasure at seeing me as we shook hands 120 feet down in the blue Aegean Sea. He waved toward what looked like a rock formation 30 feet away. Unencumbered by the lead weights, copper helmet, and patched suit that the sponge divers shared in rotation, I swam toward the outcropping. My air bubbles streamed up toward the *Mandalinci,* the three-bunk

In an Aegean cove, author Peter Throckmorton nets fish near a sunken ship. "Storehouses of history," he calls shipwrecks, that provide us with "clues to ancient ways of life."

sponge boat we called home that summer. Suddenly I realized that what looked like rocks really were amphorae, or wine jars.

They were round, and very big, like the eggs of a sea monster. Smaller jars lay among them. I swam to the edge of the pile, an oblong heap about the length of a fair-sized cargo ship, shoved my hand into the soft mud, and felt spongy, worm-eaten wood. I pulled one of the smaller jars free of the mud and tied it to a line the diver held out to me.

Before ascending, I took a long look around at what undoubtedly was the ruin of a ship. What ship was she, I wondered, out of what port? Where was she bound, carrying what cargo, and under whose hand? What adventures had she survived before the final one that brought her to the bottom in the lee of Yassi Island off southwestern Turkey?

As I swam slowly up toward the boat, I resolved to find out.

Since that summer of 1958, when I shared the sponge diver's life with Captain Kemâl Aras and the others aboard *Mandalinci*, I have seen undersea archeology develop in the Mediterranean, uncovering valuable historical remains in Old World waters.

The treasure trove of clues to ancient ways of life includes the oldest ship found so far, a 3,200-year-old Bronze Age wreck I identified off Turkey's Cape Gelidonya the following summer, with the help of some of the same men who had led me to the discoveries at Yassi Island.

Such men—the Greek, Turkish, and Italian mariners who sail these waters as their fathers and grandfathers did before them— have alerted archeologists to much of this wealth of history. Captain Aras, for instance, gave me the first hint that a Bronze Age wreck might lie off Cape Gelidonya when he told of finding bronze hatchets and knives and a pile of corroded ingots that he planned to dynamite apart and sell for scrap.

And at the turn of the century a Greek sponge boat captain named Demetrios Kon-

dos demonstrated the potential rewards awaiting undersea archeologists by raising most of a cargo of classical statuary—fourth-century B.C. bronze originals and marble copies of earlier works—from a vessel that had sunk near the tiny island of Andikíthira in the first century B.C. It was en route from islands of the eastern Aegean to Rome.

Returning from diving for sponges on the banks off North Africa, Kondos had decided to wait out a southerly gale in the Andikíthira channel between Crete and the Peloponnesus. Before moving on, he sent a diver over the side to look for sponges. Minutes later the diver was back aboard, babbling almost incoherently about horses and naked men and women he had seen below. Kondos donned a diving suit, went over the side, and soon sent up the sand-filled bronze right hand of a man.

For nine months in 1900 and 1901, under the auspices of the Greek government, Kondos and his men carefully salvaged the statues and artifacts. Their finds, including a remarkable astronomical computer, the grandfather of modern clockwork, can be seen today in the National Archaeological Museum in Athens.

While some finds have yielded immediate rewards—such as the salvage value of bronze implements and lead anchor stocks, or pieces of statuary and artifacts for display in museums—not until the early 1950's were some archeological techniques applied to Mediterranean wrecks. The first important underwater excavation began in 1952; Capt. Jacques-Yves Cousteau and archeologist Fernand Benoît spent five years excavating a Greek wine freighter that had sunk between 200 and 100 B.C. ten miles east of Marseille. In the years that followed, some land archeology techniques were adapted to sunken vessels, and new methods of plotting, photographing, lifting objects from the sea, and preserving them were introduced.

Some of these innovations came about in

Bosun Manolis Maltezos squints in the sun's glare aboard the *Archangel*, Peter Throckmorton's salvage vessel. The diver-archeologist used the ship in 1967 during excavation of a Roman wreck that an Italian fisherman led him to in the Gulf of Taranto, the instep of Italy's boot. Other Mediterranean seamen have fouled anchors on wrecks or netted antiquities at similar sites. Their advice has led archeologists to many ships. Today, modern electronic equipment helps to pinpoint buried vessels.

KIM HART

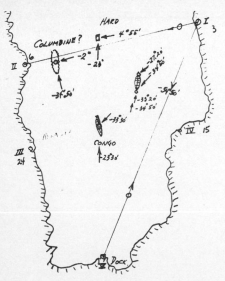

KIM HART

108

connection with wrecks I worked on beginning in 1960 with George Bass, an archeologist from the University of Pennsylvania. As a merchant seaman turned freelance writer, I had long been fascinated by the sea and the ships that sail it. Now I became interested, too, in the ships that lie clutched in its depths. Since then, I have become a specialist in surveying and studying ancient ships underwater.

In 1963 at Pórto Lóngo, a little harbor on the Greek island of Sapíentza in the Peloponnesus, we began trying to locate a wreck by pure historical research: By 1969 we were still looking for it, but with a modern tool, a sonar device invented by Dr. Harold Edgerton, professor emeritus at the Massachusetts Institute of Technology. What we learned at Pórto Lóngo about the wood-preserving qualities of mud was put to use later on the south coast of England to help evaluate the condition of the *Mary Rose,* a store of Tudor history buried in the mud off Portsmouth Harbor.

Finding ships is only part of the battle; they must be preserved and reconstructed. The Swedes pioneered in this field with the warship *Vasa,* brought up in 1961. In Cyprus, Michael and Susan Katzev have put together a fourth-century B.C. Greek merchantman that they began recovering off Kyrenia in 1968. The Dutch East Indiaman *Amsterdam,* grounded on the coast of England, still awaits salvage and preservation.

Now there are so many projects going on

Graveyard of sailing ships, the deceptively placid Greek harbor of Pórto Lóngo lies on trade routes sailed by mariners for centuries; sudden winds have sunk many ships here. Probing for wrecks, Massachusetts professor Dr. Harold Edgerton uses sonar he adapted to penetrate sediment. It pings sound waves into the bottom and records the shape of buried objects. On a map sketched by Edgerton the signals give divers targets to investigate.

in Europe, so many ships being recovered in whole or in part, that it's hard for one man to keep up with all of the developments. The men and women who worked as underwater archeologists with George and me in the 1960's have gone on to other things. In London, Joan du Plat Taylor has founded the *International Journal of Nautical Archaeology;* George Bass and Honor Frost are advisory editors. Honor is currently excavating a Phoenician wreck at Motya, Sicily. A. J. Parker, who teaches classics at the University of Bristol, is also working in Sicily, diving on Roman wrecks.

The cities from which the wrecked ships sailed have become the specialty of England's Nicholas Flemming, who has swum above a hundred ancient, submerged ports.

One of the most highly regarded members of the group is George Bass, who first dived with me in Turkish waters off Cape Gelidonya. There George and I employed such standard surface archeological techniques as surveying the site by tape and triangulation and marking artifacts with numbered plastic tags before photographing and removing them.

The two following years, excavating off Yassi Island where the amphorae had so intrigued me, he experimented with a new method of surveying—laying a grid over the wreck to provide coordinates for mapping it with stereophotography. The meticulous work of George and his crew provided some answers to the questions that had haunted me as I swam to the surface after visiting the wreck in 1958:

What ship was she? We will never know exactly, but copper and gold coins bearing the profile of the Byzantine Emperor Heraclius indicate that the vessel sank around the time of his reign, A.D. 610-641. *Where was she bound, carrying what cargo, and under whose hand?* The 900 amphorae suggest wine as the principal cargo, and possibly she traded among the islands. A steelyard,

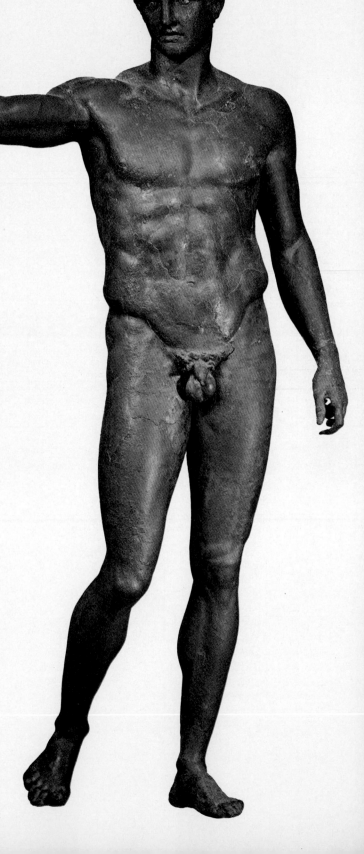

Timeless Greek statuary—fourth-century B.C. bronzes of a man and a boy—lay in the Mediterranean for 2,000 years. Antiquarians pieced together the 6½-foot-tall man from fragments; some authorities see him as an athlete who has just thrown a ball. Sponge divers found him 180 feet down off the tiny Greek island of Andikíthira in 1900. The four-foot-tall youth, recovered nearly intact off the coast of Marathon in 1925, gazes at his left palm; a hole there indicates he may have once held an object—perhaps a bowl into which he poured water, or carried picked fruit.

used by Byzantine traders for weighing goods, answered the final part of that question with a simple inscription: "George the Elder, Senior Sea Captain."

One man who has proved invaluable to many of us is Dr. Edgerton, whose sonar has helped find buried wrecks that sponge divers and trawler crews could not tell us about. I first worked with the genial professor at Pórto Lóngo.

In London, while pursuing one of my hobbies—the study of Lord Nelson's Navy —I had come across an account of the loss of the small British warship *Columbine*, sunk in Pórto Lóngo in 1824.

Unlike the dozens of ancient wrecks found in that area, the *Columbine* had her loss fully documented, along with the court-martial of her captain. I am fascinated by such old reports for the clues they give as to why presumably able captains lost their ships. And I believe that charting recorded wrecks should lead to much older ships that sank in the same places for similar reasons.

So one day in 1963, armed with the court-martial transcript and the *Columbine*'s log, which her captain had saved, I conned a launch into Pórto Lóngo, trying to watch the compass and decipher the copperplate script: "At 6 a.m. a heavy squall from the SE parted the Small Bower (bow anchor). Let go the best bower. Parted the warp (anchor rope). Ship still driving. . . ."

Our boatman kicked the tiller a bit, and we drifted into the anchorage on the same heading as the strong southeasterly wind that had blown 139 years before.

"At 6:30 the ship struck very hard . . . saw several pieces of the keel and sternpost floating . . . three feet of water in the hold

Eyes of white stone gaze from a bronze head of the fourth century B.C. The world's first marine archeological expedition, sponsored by the Greek government, recovered the head and fragments of limbs in 1900 off Andikíthira.

and rapidly increasing. . . ." We came under a rocky cliff that seemed like the place where *Columbine* had struck and "lurched to seaward, filling fast and drawing into deep water." In a few minutes we had our air tanks on and were in the clear water. We knew that *Columbine* should lie six fathoms deep within a couple of hundred feet of where we were moored. In an hour we were back in the boat, chilled to the bone. We had found no sign of the *Columbine.*

However, we had found the remains of a large wooden ship that our boatman identified as the *Heraclea,* a Greek schooner carrying a cargo of wheat when German dive bombers sank her in 1940. Working as a helmet diver after the war, he had helped take out her engine. Now she was settling into the muddy bottom of the harbor, showing us what probably had happened to *Columbine.*

We dived with scuba, using iron rods to probe the soft mud. On the third day a diver surfaced with a whoop, waving a bit of worm-eaten wood; he had found the remains of a ship under about six inches of mud.

We believed this wooden wreck was the *Columbine,* and were delighted when an airlift vacuumed off enough mud to reveal gunports. But a discrepancy nagged me: The timbers were softwood, not the good English oak common to most early 19th-century British warships. We decided to dig a trench across the wreck and draw a cross-section of the existing timbers.

We dug. We drew. We dug some more. We had photostats of the drawings from which *Columbine* was built, and we spent our evenings in long discussions over what might match with what. After a month of digging we had the answer. Nothing matched. The ship was not the *Columbine.*

By this time, most of the nearby village of Methóni was involved with the question: If the ship wasn't *Columbine,* what was she?

One evening a local lawyer, Constantine Vassopoulos, appeared at our camp with a possible answer. The year before, searching through public records for a land title, he had come across the story of an Austrian brig, a square-rigger lost in 1860. That, Mr. Vassopoulos speculated, must be the ship we had found.

The brig had sunk in January, in a southerly gale, and so had *Columbine,* a generation earlier. We know almost nothing about the brig, except that she existed. Yet the final story of both ships is undoubtedly the same.

After my initial search proved unsuccessful, I let six years pass before returning to Pórto Lóngo—this time accompanied by Dr. Edgerton and his sonar. From the accumulated experiences of all who had worked on wrecks since 1952, I knew that ships sunk and buried in mud or sand were the best protected from destructive shipworms.

But while divers probed the mud to find a ship hidden in it somewhere, time and money could run out. About 1965 a quick way of seeing through sediment appeared in Dr. Edgerton's two sonars, the side-scanner and the sub-bottom "pinger." The side-scanning sonar draws a continuous profile of the sea bed as it would appear if you were standing on the bottom; the pinger penetrates sediment and records the shape of anything solid in it.

In 1969 and 1970, we probed Pórto Lóngo with Dr. Edgerton's sonar and with magnetometers. In the end we had a map of the subsurface of the harbor bottom pinpointing half a dozen modern wrecks and heaps of ballast stones invisible to divers.

The readings on Dr. Edgerton's instrument became meaningful when we had explored the wrecks with an airlift. We learned that the sonar could not detect waterlogged wood, but that it could locate ballast stones. And it made the clearest graphs of all when the mud was filled with gas, like that from

Heraclea's rotting wheat or from a cargo of raisins that went down in another ship, the *Congo*, in the 1870's.

Although we never did find *Columbine* in Pórto Lóngo, we did learn more about how to interpret the lines on sonar graphs and what sonar can "see." As has happened time and again in undersea archeology, learning gained on one wreck helps solve puzzles on another.

My experiences with ships lying in mud, for example, helped historian Alexander McKee make a decision about whether to continue his search for a sunken Tudor ship.

The story he was investigating began on a sunny July day in 1545. A French fleet of 235 ships was attacking Portsmouth Harbor. The carrack *Mary Rose*, Captain Roger Grenville commanding—with Sir George Carew, Vice Admiral of the English fleet aboard—had sallied out to engage them.

Henry VIII, who 35 years earlier had ordered the construction of the *Mary Rose* and designated her his flagship, watched as mishandled sails capsized the vessel. Watching, too, was Admiral Carew's wife; of perhaps as many as 700 aboard, fewer than 40 survived, and neither her husband nor the captain was among them.

Mary Rose slumbered for three centuries, one of the thousands of wrecks littering Portsmouth's muddy anchorage. Then in 1836 pioneer diver John Deane accidentally found the carrack while salvaging a ship sunk in the 1780's. Deane recovered a few of her 91 cannon and then quit, as the ship was completely covered with mud. Details of her construction are almost unknown, but she was the first English warship to carry complete batteries of siege artillery

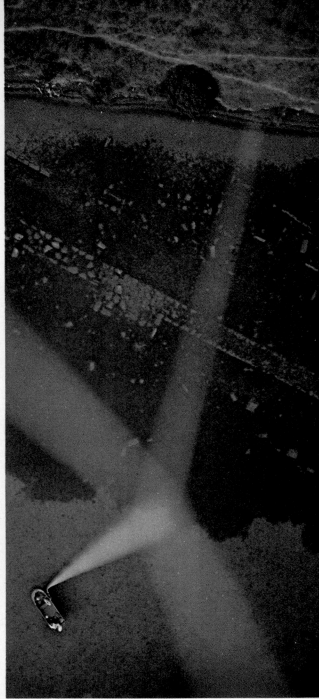

WHITTLESEY FOUNDATION

Ruins of the sunken Greek city of Halieis appear clearly when photographed by a balloon-borne camera tethered to a boat. Residents abandoned the port in 300 B.C.; hundreds of such sites rim the Mediterranean.

on the main deck—an important innovation.

By 1967, aided by an old wreck chart, Alexander McKee believed he had found the general location of *Mary Rose*. But was she in good enough condition to be raised? Enthusiastic about the possibilities of the project, I readily accepted an invitation to visit the location and give my opinion.

I was less enthusiastic when I reached the site, the Solent strait a mile off King Henry's Southsea Castle. Chill gusts of rain slapped at the launch's deckhouse. I wondered if there wasn't a way I could get out of diving into those steep gray seas.

I smiled feebly at Alexander as I struggled into a borrowed wet suit, which promptly split at the shoulders. I hurried to get the tank on, and dropped over the side to meet Alexander for an equipment check ten feet down. I could see as far as my elbow, more or less. I gave Alexander the okay signal and dived after him into the darkness. We hit bottom 40 feet down in a cloud of mud and, at that moment, my borrowed suit split across the back.

As the icy water of the estuary wrapped a numbing belt around my middle, I knelt on the bottom and concentrated. Ah, yes! Mud! I was to judge the mud. There it stretched, all around us, blackish gray, covered with concretions and mollusks, riddled with crab holes. I stuck an arm elbow-deep into the gelatinous stuff as Alexander passed me a steel probe. Lovely mud, I thought. Softer, finer, stickier than any I had ever encountered. Here the mud would surely have protected the *Mary Rose* as thoroughly as it had preserved the Austrian brig in Pórto Lóngo.

Back aboard the launch, we talked about the knowledge of wooden wrecks gained from Pórto Lóngo: how they slid under the mud, how safe they remained once covered. I told him I thought at least half of the *Mary Rose* would be intact. She would probably have filled quickly; even paper might be preserved. Digging through the mud would

be cheap and easy, if he had good equipment. Alexander was encouraged to continue.

In October 1967, during a demonstration of Dr. Edgerton's new side-scanning sonar, an anomaly showed up on the graphs—a disturbance on the surface of the sea bed—that he hoped was the *Mary Rose*. That was enough for him to obtain a lease on the site from the government.

Dr. Edgerton carried out a full-scale pinger survey in July 1968, and resulting charts puzzled everybody. Did certain lines indicate that internal timbers had collapsed? It took more than a year for divers to trench deep enough to suspect that the lines were caused by masses of seaweed in the wreck. At about the same time, in Pórto Lóngo, I was discovering the rotting raisins and wheat that had shown up so distinctly on the sonar charts. Alexander and I compared notes in 1971: Sonar reacts more strongly to gas-filled mud than to rock.

All these years Alexander had little money, but he did have the frequent loan of boats and equipment, and the help of volunteers from diving clubs. Margaret Rule, an archeologist who had helped from the start, became so interested she learned to dive.

Alexander and his friends used airlifts to dig trenches across the site. The trial trench dug in 1969 and 1970 finally extended 90 feet—a 12-foot chasm 25 feet wide. More were dug the next season.

Working in cold tidal waters amid nightmarish landscapes of sediment clouds, they at last found a Tudor cannon, and in 1971 uncovered a portion of the hull. Finally, in May 1972, they came upon the sterncastle.

The *Mary Rose* is nearly intact, lying with a slight starboard list. Bones of lost sailors litter the wreckage. The English oak of the great hull, solidly held together by wooden pegs, seems nearly as strong as when she sank 430 years ago. But the volunteers have much to do before they can raise the *Mary Rose*. Not the least is providing for the

116

preservation of this important Tudor vessel.

Techniques for preserving salvaged ships have come a long way since that day in 1961 when the waters of Stockholm Harbor roiled as the Swedish warship *Vasa* rose from the bottom where she had lain for 333 years. The *Vasa*, which sank intact less than a mile from shore during the first hour of her maiden voyage, came up intact, as did even the butter and rum in the ship's stores.

To raise the ship, which had been located by amateur marine archeologist Anders Franzén, salvagers burrowed six tunnels in the mud beneath her, threaded cables through them, and attached the lines to pontoons on the surface. The pontoons were partially filled with water to lower them, and the cables were tightened; then the water was pumped out, and the pontoons floated higher, raising *Vasa* slightly. The process was repeated 18 times over a two-year period, with *Vasa* moving gradually from a depth of 100 feet to 50 feet on the harbor bottom before a final lift brought her to the surface.

Technical problems with the *Vasa* were many. The hard oak planks of the ship, if left to dry, would have contracted and warped. Lars Barkman, the project's conservationist, developed a preservation process that is being used all over the world today; it depends on a particular kind of polyethylene glycol that replaces water in the wood's cell walls. The replacement material allows the wood to retain its original shape. After more than a decade of treatments the *Vasa* is now stable.

More than 400,000 people a year view the *Vasa*, the central attraction in a temporary

Cofferdam of sandbags and concrete surrounds an excavation—pumped dry during digging—in the Grecian harbor of Kenchreai. Opaque glass mosaics of the Greek philosopher Plato and birds of the Nile Valley survived an earthquake in A.D. 375 that drowned the site.

museum built especially for her. Such museums, I hope, can provide future generations with a link to the great age of sail.

Now the Dutch are trying to raise enough money to salvage and display the merchantman *Amsterdam* in her namesake city. This spectacular Dutch wreck turned up near Hastings, England, in 1969, when local workmen laying a sewage pipe dug up more than 1,000 objects, ranging from bronze cannon to horn combs, from bottles of soured wine to brushes with leather handles.

England's Committee for Nautical Archaeology managed to prevent souvenir hunters and amateur archeologists from destroying the wreck. Study of the objects revealed the nationality of the ship and the fact that she had sunk in the middle of the 18th century. A search of Dutch archives positively identified her as the *Amsterdam*, an East Indiaman bound for Java when she ran aground in quicksand in January 1749. Nearly the whole ship and her cargo remain intact, and I hope that someday she can return to Amsterdam for exhibition in her own museum.

In Germany, a wreck raised during the dredging of Bremen Harbor in 1962 will be reconstructed and displayed in an aquarium filled with polyethylene glycol until it is stable—perhaps as long as 25 years. The ship is a good example of a small cog, one of the host of merchant vessels which made the Hanseatic League of German cities a commercial power in medieval times.

The Baltic contains dozens, perhaps hundreds, of ships as intact as the *Vasa*; in this northern sea the water is neither salty nor warm enough to support shipworms. Wrecks from Russo-Swedish sea battles of 1780 and 1790 were being mined a century later, their sea-blackened oak salvaged for furniture and house interiors. But only with the advent of scuba in the 1950's did the

(Continued on page 124)

Trayload of wood planks emerges from a bath in polyethylene glycol. The solution permeates waterlogged ship's timbers and prevents their shrinkage in open air. Riddled by marine borers (above), the keel of the oldest Greek vessel ever recovered awaits treatment. Damage by shipworms may have contributed to the merchantman's sinking at the end of the fourth century B.C. off Kyrenia, Cyprus. Surveyed in 1967, the wreck yielded five tons of worm-riddled pine during two years of recovery work. Reconstructing a frame, excavation director Michael Katzev (below, at right) and two colleagues follow full-scale drawings to reshape the piece to its original curvature.

Overleaf: Framework of plastic pipe divides the 40-foot skeleton of the Kyrenia wreck into sections for excavation. Photographs of the site reveal how the wine freighter settled into the sand and broke apart. At a depth of 90 feet, divers airlift silt that accumulated on the vessel during 22 centuries.

Worm-eaten timbers of the Greek
merchantman rest on a scaffold in
Kyrenia Castle, built by the Crusaders
in the 13th century. In the summer of
1974 the Cypriot government opened
a museum in part of the structure,
displaying the vessel and her cargo
of millstones, wine jars, pottery, and
thousands of almonds. Air condition-
ing prevents the preservative from
deteriorating; masking-tape labels
stuck to each part aid in the
reconstruction of the ship. A member
of the restoration team drills holes
for wiring a frame to the planking.

JONATHAN BLAIR

existence of these wrecks become widely known; much material was salvaged, almost all of it improperly, and as a result it became junk as soon as it was raised. The delicately carved figurehead of the *St. Nikolai,* a Russian frigate sunk in 1789 and discovered in 1948, shrank into an unrecognizable caricature of itself.

Scuba not only opened the bottom of the sea to scholars, but also to scavengers. In the early 1960's, as George Bass of the University of Pennsylvania Museum was creating an underwater archeology department that lasted until he left in 1973, sport divers off France, Italy, and elsewhere in the Mediterranean were scooping up amphorae for souvenirs. I think it's fair to say that they have destroyed all the easy wrecks off the southern coast of France, and many off Italy and Greece.

Fortunately, some wrecks lie in spots remote enough to escape sport divers. George Bass, now head of the American Institute of Nautical Archeology based in Philadelphia, is excited about one such ship, an Iron Age coastal freighter. A sponge diver chanced upon the vessel's remains in 1967 and told George, who believes he stands a good chance of salvaging some of her timbers and gaining an insight into ship construction of the seventh century B.C. To prevent sport divers from beating him to her, he has been keeping the exact location a secret, saying only that she sank off Turkey.

Perhaps looting will decline as steps are taken to protect wrecks. Already there are encouraging signs: Laws to protect sites are being passed in many European countries,

Corroded and encrusted by centuries in the sea off Andikíthira, this geared instrument once helped Greek astronomers calculate the positions of the sun, moon, and stars. Yale science historian Derek de Solla Price calls it the most complex mechanical device surviving from Classical times.

and underwater archeological institutes have been established. The Danes have been especially successful, and the Viking Ships Museum at Roskilde has become a center for the survey, study, and preservation of Denmark's maritime past and traditions.

In Greece, a group of divers and archeologists has set up the Hellenic Institute of Marine Archaeology. Great Britain has a maritime archeological institute associated with St. Andrews University.

One of the principals in the struggle to educate divers is Alan Bax, who founded the Fort Bovisand Diving Center in Plymouth, England, when he retired from the British Royal Navy. Alan offers courses in nautical archeology and holds classes for his students on the wrecks which stud the bottom near Plymouth.

Alan and I visited one of his underwater classrooms, H.M.S. *Coronation*, a 90-gun vessel built in 1685 and wrecked seven years later. The water was icy but quite clear, and we easily found a mass of cannon 30 feet down in a grove of kelp.

As we surfaced, divers in a rubber raft arrived—souvenir hunters of the sort that Alan hopes to reach and influence with his underwater archeology classes.

Many sunken wrecks are now coming to light by various means: The draining of the Zuider Zee in the Netherlands, for instance, exposed several hundred ships, most from the 18th and 19th centuries. And throughout Europe a surprising number of vessels have been found buried *inland* from the sea, the result of changing coastlines.

Perfection of sonar and magnetometer survey techniques promises to reveal more sites. I can speak with confidence, for in 1971 Dr. Edgerton and I began searching for a naval battle site and have located many possible wrecks. Although five are less than 100 years old, we presume most of the others sank on October 7, 1571, when Turks and Christians clashed in the Battle of Lepanto.

The exact site of this battle is uncertain, but the general area has always been known: the Bay of Patras, below the mouth of the Akhelóös River on the west coast of Greece. There, about 100,000 Turks in some 400 vessels fought 60,000 Christians in 300 vessels. More than 15,000 men and dozens of ships—many of them galleys rowed by slaves—went to the bottom. The victory, albeit an expensive one, belonged to the Christian fleet put together by Spain, Venice, and the Papal States.

In five extended cruises aboard my schooner *Stormie Seas* we plotted 1,000 miles of survey tracks in a 75-square-mile area, towing one short-pulse sonar to probe the muddy bottom directly beneath us, one narrow-beam side-scanning sonar, and a magnetometer to detect the light artillery, chains, anchors, and grappling hooks that many of the ships carried. The result is a chart of an area sprinkled with small targets for future exploration, but most are in depths dangerous for scuba divers—150 to 200 feet. Mud should have preserved the timbers. We believe many could be remnants of the Lepanto fleets. Others may be really ancient vessels—Phoenician, Greek, Roman, or even Bronze Age.

Wherever they sank, and for whatever reasons, most took men down with them —captains, sailors, passengers, warriors, slaves. For me, the discovery of wrecks presents the opportunity to learn not only about the ships, but also to speculate about the men who sailed in them.

Theorizing about a Turkish viceroy's battle plans and general orders to his captains, or why a Roman captain tried to run through Andikíthira channel on a stormy night, may not be wholly scientific, but I find such efforts satisfying. And I like to think that George the Elder, and all seafarers too pressed to write that final episode in their logs, wouldn't mind if one of their own spends his time trying to do it for them.

Preserved between sheets of clear plastic, remnants of sails
from the Swedish warship *Vasa* hang in the Stockholm museum
dedicated to her restoration. The vessel capsized and sank in
the city harbor when hit by a sudden gust of wind on her
maiden voyage in 1628. In 1961 salvors slowly raised her
with lifting pontoons attached to cables beneath her hull.
Below, in the museum's display gallery, the main deck
undergoes a spraying with polyethylene glycol. Above, gold
leaf flecks the sea-blackened oak head of a Roman
warrior, one of *Vasa*'s more than 700 carvings.

Hull and cargo buried beneath fine sand, the Dutch East Indiaman *Amsterdam* awaits excavation and restoration. On her maiden voyage in 1749 she lost her rudder and ran aground in quicksand near Hastings, England; within two months she had sunk 30 feet. In 1969 workmen laying a sewage pipe noticed the wreck nearby and dug into it. Realizing its importance, they informed archeologists. As part of her cargo, the ship carried 12,000 bottles of wine, like the one at right. Of 60 bottles recovered in early digging, only five survive, all soured; souvenir hunters made off with the rest. Other finds include bronze gauges for cannonballs and iron shot, and a reamer for cleaning cannon touchholes; an intricately carved ivory fan; and wooden pulley blocks and rope from the ship's stores. Pewter spoons, one bearing a roughly scratched image of a ship, belonged to passengers and crew. Salvors hope to complete excavation of the ship by 1975, *Amsterdam*'s 700th anniversary.

PETER MARSDEN
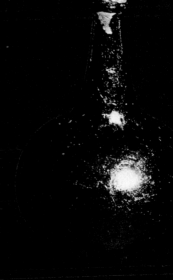

Diving for New World Wrecks

6

BY MENDEL PETERSON

LISTING about 15 degrees to starboard, a lady entered my office in the Smithsonian Institution one day in 1951, carrying a small but heavy canvas bag. She was Jane Crile, from Cleveland, and as she emptied the bag onto my desk, my historian's nose began to twitch. Among the objects, which Jane said had been taken from several shipwreck sites on the coral reefs of the Florida Keys, was a cast bronze breechblock for a small swivel gun. An iron cannonball in her bag bore an impression shaped like an

In murky Jamaican waters, archeologist-diver Robert Marx retrieves a wine bottle. Such men add to our store of historical artifacts as they probe the New World's maritime past.

arrowhead. I recognized it as the broad arrow—royal property mark of England.

What ships had they come from, I wondered, and what other historically valuable artifacts might still lie in the wreckage? With delight I accepted Jane's invitation to join her and her husband, surgeon George Crile, Jr., and half a dozen friends, to explore further the wrecks they had visited. I didn't know it then, but our trip would mark the beginnings of archeological work in North American waters—and the beginning of my career in that field.

In the spring I joined the Criles in Miami, and we drove from there to the Keys. I rode in an open touring car loaded with supplies. A 20-foot section of black hose we had bought in a junkyard stuck out of the trunk like a captured python. We would use it as an airlift, to vacuum sand from the bottom.

At Marathon, about halfway down the Keys, we met Bill Thompson, who had shown the Criles one of the wrecks we were going to explore. Also present was Art McKee, who had found some others. And I learned that engineer-inventor Ed Link would visit us in his racing yawl Blue Heron.

There in Marathon I spent a day in a shallow swimming pool learning to use a diving mask and to breathe compressed air pumped from the surface. It seemed easy enough. But on the wreck site, when I jumped in—cotton work clothes and all—I learned that the open sea is not like a swimming pool. I felt very insecure in that limitless, liquid world, and had to make a sustained mental effort for two years before I overcame my fear and became a competent diver.

Although I was not a trained archeologist, I knew the essential techniques, and that digging on a site must be done carefully and precisely. When I went over the side, weighted with a hammer and stakes, I had in my hip pocket a large roll of red tape that I planned to use in laying out a grid.

Underwater I found myself in a narrow valley. A jumble of coral chunks, covered by living corals and soft sea plants, lay between walls that climbed sharply toward the surface. I almost blew off my mask laughing as I visualized myself trying to lay a neat red-tape grid in such a place.

For three weeks we collected artifacts from the mysterious sunken ship—cannonballs, ship fittings, bolts that held the great timbers together, the ring of one of her anchors, wrought-iron chain plates that held deadeyes for lines running to the masts.

Cast-iron bars, each some three feet long and six inches broad, were used as ballast in the early 18th century by the British Navy. A crowned rose marking on a cannon indicated that the gun was probably cast before 1727, for that symbol was not commonly used after the death of George I. Since an iron cannon would only last about 30 years in the abrasive salt air at sea, ship and gun must have gone down before 1750.

As Ed Link and I sat talking one afternoon, when wind and rain kept us ashore, we spoke of the history that lies buried under the sea, about the type of salvage vessel best suited to finding and studying the sunken relics, about the airlift, waterjet, lifting equipment, and the great potential of scuba, then just coming on the market. I could almost see Ed shifting his hobby from ocean racing to underwater exploration. That conversation proved to be a turning point in both our lives.

Before the expedition wound up, Bill Thompson guided us to a 17th-century Dutch warship, then to a merchant vessel. I decided it must have been a slave trader, for we found elephant tusks, brass pans for feeding the slaves, and remains of light, poor-quality muskets often traded for slaves.

Back in Washington, I began an attempt to identify the English ship we had explored. It lay on the reef of Looe Key, and in a reference book I eventually found an entry—"Looe, forty guns, lost in America, 1743."

ROBERT MARX

"Cannon literally point to a site," says the author, "...often the only visible evidence that a wreck exists." Here a diver clears sand from a ten-foot bronze gun off the *Maravillas*, a treasure galleon that sank one night in 1656 after ramming another ship as the convoy veered seaward to avoid shallow water. Robert Marx found the ship in 1972 north of the Bahamas 35 feet down. Cannon often help salvors establish the size, type, and approximate date of a ship—but not necessarily its nationality; ships often carried guns from the foundries of many countries. Sometimes, too, mariners used a jumble of old cannon as ballast, and to lighten a grounded ship they threw cannon overboard, creating false sites that occasionally mislead treasure hunters. Minted in Mexico City, a piece of eight embedded in coral also came from the *Maravillas*.

The Key had taken its name from the ship.

In London the next year, I dug out all the ship's surviving documents and published an account of its last voyage, which ended when an unexpected current swept it onto the reef. Research on the *Looe* and my first diving experience decided my future. It would be under the sea.

At the same time, Ed sold *Blue Heron* and bought a shrimp trawler, renamed *Sea Diver,* which he fitted out for undersea exploration. Also, he generously established a fund for marine archeology at the Smithsonian, thus permitting me to carry on a program there and to join him in field work.

In the summer of 1952 we explored shipwrecks from Looe Key south to Key West. We learned much about the appearance of wrecks, the condition of materials found on them, and especially the importance of cannon in finding and identifying ships. Cannon literally point to a site, and are often the only visible evidence that a wreck exists. Their markings frequently—but not always —indicate a ship's nationality and period.

The next two summers we worked in the central Keys. I had improved my diving by training with scuba in the Navy's Experimental Diving Unit tanks in Washington; now I could work as well in water as I could on land. I still prefer the mask and hose for hard work in shallow water. Without the bother of coming up for a new tank of air, I can work three-hour stretches. I did that on the *Rui,* for example, the wreck Art

JENIFER MARX; DRAWING BY PETER COPELAND

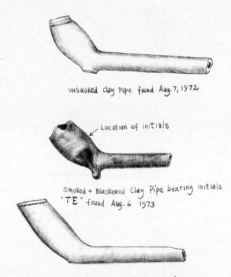

UNSMOKED clay Pipe found Aug. 7, 1972

← Location of initials

Smoked + Blackened Clay Pipe bearing initials "TE" found Aug. 6 1973

unsmoked Clay Pipe found Aug 5, 1972

English clay pipes recovered by Robert Marx number among some 12,000 he found in Port Royal, a Jamaican city tumbled into the sea by an earthquake in 1692. Pipe smoking spread from America to Europe in the 1500's. Through the years, stem length and bowl design changed; bowls got larger as tobacco became cheaper. The differences help archeologists date sites. Pipes from a wreck off Bermuda (right) reveal styles of the 1640-1690 period.

McKee had been excavating for several years. Shimmering white ballast stones covered an area 150 feet long and 70 feet wide on the sand floor. On the stones, sand-encrusted cannon lay where they had fallen as the ship was pounded to pieces by a hurricane. No timber was visible, but we later found a portion of the ship's hull buried beneath the ballast.

One day I was working my way with a water jet under the edge of the timbers, finding ship fittings and nails. I was so absorbed in my work that an hour passed as I tunneled deeper and deeper beneath the structure, collecting hundreds of pounds of sand-encrusted iron. Suddenly it occurred to me that over my head there were several tons of stone ballast sitting on rotten timbers. Worse yet, I was diving alone. I came out fast. If the timbers had collapsed I would have been trapped or crushed. My air bubbles would have continued breaking the surface above me, and my companions on the *Sea Diver* would never have suspected that I was in trouble. Back on deck I felt lucky that my only injury was a patch of skin missing from my left arm. I had sandblasted it off with the water jet in the excitement of pulling objects out of the sand.

During that month in 1953 Art gave me a graduate course in underwater observation. Ed Link, at the same time, taught an advanced course in engineering, for as each problem in salvaging arose, he furnished the solution. He had designed and brought with him a new airlift and an improved detector; both worked beautifully. With the new detector we found in the great pile of ballast stones deposits of iron—a cluster of cannon, swords, flintlock muskets, pistols, and a 12-inch pewter bowl. Sailors probably once gathered around and ate from it with their knives.

Most interesting to me were small cast iron grenades still filled with black powder; their wooden fuses had rotted away. We also found ship fittings, nails, bolts, and iron straps that had once held the timbers together. These artifacts, encrusted together in clumps sometimes weighing more than a hundred pounds, resembled the fantasies of a demented sculptor. The objects our hammers released from the clumps gave us a picture of an early 18th-century treasure ship that no document could ever provide.

For the 1954 season, we borrowed a magnetometer from the U. S. Navy. We also had a map I had located in the Library of Congress that showed sites where ships of the Spanish treasure fleet went down off Florida in 1733. With it and the magnetometer we found in just a few days several wrecks I think were from that fleet, but we had no time to explore them.

In three years our methods of underwater exploration had become much more sophisticated. We had progressed from a small jury-rigged fishing boat and racing yawl to Ed's recovery vessel, equipped with the latest air pumps, diving equipment, lifting gear, and electronic detectors.

For 2½ months in 1955 I cruised with Ed in the Caribbean, trying to trace the route of Columbus—with his journal as a guide. Later we looked for remains of the ship William Phips had salvaged 250 years earlier, but I'm not certain we located them.

But the greatest treasure we found that summer was the reef itself. "Lily pads" we had seen from a plane at 5,000 feet turned out to be tops of great trees of staghorn coral, flaring outward from trunks up to ten feet in diameter. At a depth of 75 feet, with bright sun filtering down through the aquamarine water, I thought of a towering cathedral, for the coral heads intersected to form Gothic arches overhead. It was the most breathtaking sight I had ever seen.

In 1956 Ed and I turned to the great archeological site of Port Royal, on the south side of Jamaica. That sunken city contains the greatest deposit of 17th-century artifacts in existence. After the English captured this island from Spain in 1655, buccaneers went out from Port Royal to raid and pillage the Spanish possessions in the Caribbean, and many merchant ships called at the port. Merchants built warehouses to store pirate loot, merchandise, and ships' supplies— and waterfront values rose higher than London's. Port Royal became noted for wholesale sin. A minister called it "the wickedest city on Earth." On June 7, 1692, an earthquake "heaved and swelled like rolling billows," one account says, and great fissures split the waterfront. Within two minutes, two-thirds of the city and 2,000 people slid into the sea. Survivors fished out some articles, but the bulk remained trapped beneath fallen buildings.

For two weeks our expedition worked the site, recovering a modest collection of artifacts; surveys and charts prepared by Ed laid the groundwork for a full-scale effort three years later.

By then Ed had *Sea Diver II*, steel-hulled, 250 tons, equipped with the latest diving and detection gear. In a large chamber aft, divers could come and go by a ladder through an opening in the stern.

Among others, our crew in 1959 included an expert team of Navy divers and Captain P.V.H. Weems of Annapolis, noted pioneer in air navigation. Financed by a grant from the National Geographic Society, we probed, pumped, and dug in the mud-blackened waters of the harbor the entire summer. We found the rough brick walls of Fort James, nearly whole, standing in some 30 feet of

Adapting land techniques to underwater archeology, diver Pat Witters sketches a fragment of a ship's timber off Bermuda. Coral-encrusted iron and a brick hold her down; air comes from a topside compressor. Such sketches done on the bottom provide guides for scale drawings of the original ship.

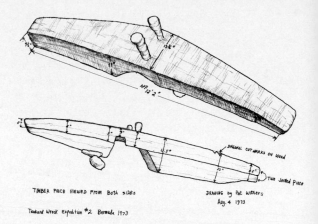

black water, the battlements cracked, the cannon lying where they had fallen from their carriages. With the pumps shut down for a couple of days, the water cleared enough that Luis Marden of the National Geographic Society could turn on his lights and make the first underwater photographs of the ruins.

As week followed week, the collection grew, although divers lying in utter blackness on the mud bottom had to feel through the muck to find artifacts and trace the foundations of buildings. A major accomplishment was the preparation of a chart of the old city based on a survey by Ed and Captain Weems.

As the work drew to a close we gathered all the finds together on a dock. Every phase of life in old Port Royal was represented: wine bottles and pipes from taverns; ship fittings, an unused yard, and a wrought iron swivel gun from the chandler's shop; cookpots—one containing bones from a stew—a brass skimmer, pottery bowls and dishes, pewter spoons, a porridge bowl and platter from the kitchen; a weight from the market; a brass candlestick from the bedchamber. Most dramatic of all was a handless watch. At the request of C. Bernard Lewis, Director of the Institute of Jamaica, a dentist took an X-ray photograph of the watch; it showed traces of what may have been the missing hands positioned at 11:43, about the time the disaster is known to have occurred.

The collection has now found a permanent home in the Institute of Jamaica, Kingston. Thousands of artifacts were added ten years later when Robert Marx, working as an archeologist for the government of Jamaica, spent three years systematically searching 1½ square miles of shallow waters. The list of his finds is astounding. They represent a priceless resource for students of English colonial social and economic history, and are a tribute to Marx's energy and ingenuity.

By 1960, I had spent nine years participating in archeological exploration in the sea. In the Mediterranean, Capt. Jacques-Yves Cousteau had excavated a wreck near Marseille in 1952-1957, and in 1960 George Bass began directing underwater digs on ancient ships off Turkey. But in all of the United States, the Smithsonian Institution's modest program was the sole government-sponsored activity in historical exploration under the sea.

Now the reefs of Bermuda and their hundreds of wrecks beckoned me. Beginning in 1960, I shared leadership of several expeditions with Bermudian Teddy Tucker. Treacherous reefs extend outward from his islands as much as ten miles, and locked in their sand and coral lies a vast archive of history. When I appraised Teddy's treasure in 1955, I saw at the same time the unique collection of artifacts he had recovered.

There were swords, a sand-encrusted steel breastplate, and a dagger. As a naval historian, I was excited to see several graceful brass dividers, used to chart a ship's progress, and a finely made brass case for an hourglass. Fragments of Chinese porcelain reminded us of the importance of the galleons that brought Oriental goods across the Pacific to Mexico for transshipment to Spain. Leg irons told of cruel discipline for the crew. A pewter cylinder was an apparatus for administering enemas.

Black wooden bows and an enormous carved club proved to be unique artifacts: Early engravings show Carib Indian chiefs carrying such clubs as their staffs of office. For these ethnological treasures we can thank some anonymous Spaniard who was taking home souvenirs from the New World.

During ten years of annual expeditions Teddy and I—aided by my assistant at the Smithsonian, Alan Albright, and various divers—studied numerous sites of all periods, and tested methods of search, underwater surveying and recovery, and

PAN AMERICAN PHOTO SERVICE, INC.;
CHARLES H. BAKER, JR. (TOP AND RIGHT)

techniques for preserving finds. On the early-17th-century *San Antonio* we had our first look at a cargo of New World products.

There were fragments of pottery stained with red dye made from the crushed bodies of cochineal insects. Other fragments were covered with resin of balsam, valued as a medical dressing. We found bits of tortoiseshell, worth its weight in gold in the 17th century for use in ornaments and jewelry, and pearl shell from Margarita Island, off Venezuela. Short logs proved to be lignum vitae, that indestructible wood prized in Spain for ornaments and for pulleys in ships' blocks.

One day I dived with Teddy down to a lump of blue material bordered with rotten wood—a chest of indigo. Incredibly, impressions of the packers' fingerprints were still visible on the small balls of dyestuff.

And Teddy found treasure. Once, while I was recovering wrought-iron nails, Teddy reached into the sand right under my nose and pulled out a gold ring set with a perfect one-carat emerald. Now and again he picked up gold nuggets, a gold button set with crystal, small gold filigree buttons or studs, and gold earrings. Coins he found were the first proof I had seen that the Spaniards sometimes shipped them hidden in jars of pitch. He picked up enough gold links to make about 20 feet of chain. I found that 42 links exactly equaled the weight of one escudo. No doubt the chain was carried as a handy form of money: Its owner had only to wrench off a number of links of known value to make a purchase.

Another treasure was found by accident. The Smithsonian had obtained a quantity of sand-encrusted scrap iron for cleaning and preservation, and in a mass clinging to a cannonball was a gold ring set with an emerald, a garnet, and quartz. An inscription in archaic Spanish read, "I am yours and always will be." Since we had agreed to return anything valuable, I gave the ring to Bermuda's Governor, Sir Julian Gascoigne.

I had met this delightful man the summer before. An expedition sponsor had brought a large helium balloon to Bermuda to use in looking for wreck sites. When Sir Julian arrived to join our motley group on Teddy's rather ragged workboat, he was impeccably dressed in white trousers, fine sport shirt, and straw skimmer. Accepting our invitation to be the first to ascend, he sat down in a chair rigged beneath the great billowing red and white bag. To our horror, when we pushed the seat over the rail, the chair and Governor Gascoigne sank slowly into the water. All the time he sat quietly puffing on his pipe. Hoisted dripping back aboard, he remarked, "It is a case of too little gas and too much Gascoigne," thereby adding an immortal phrase to marine archeology.

Like Sir Julian, I was a little too much for the balloon. But the bantamweights rose up to 250 feet, gently drifted along, and spotted several important sites.

When the 1964 season opened, Alan Albright and I arrived in Bermuda with tools devised and constructed in the Smithsonian's workshops. One was quite efficient for systematically measuring ships' timbers. Another, a brass plane table, we used to take sightings and to lay out our cardinal points for measuring. Our work on an unusually large number of timbers from one wreck gave us experience that later proved valuable for identifying the English *Warwick*, sunk in a hurricane in 1619.

Abandoned barge *Charles W. Baird* sits on shallow Carysfort Reef off Key Largo, Florida. Salvors grounded the ship here in 1938 to use as a shelter while diving on H.M.S. *Winchester,* lost in 1695. Hampered by their heavy gear, helmet-and-hose divers worked the site 1½ years, recovering cannon and ship fittings. Protected by a growth of coral, fragments of a crew member's prayer book survived 245 years in the sea.

Ruthless in the pursuit of wealth, a Spanish merchant thrashes a fallen Indian porter carrying goods from Acapulco to Mexico City, on the "China Road," part of a route across 325 miles of Mexico to Veracruz. Along the road, little more than a mule path, flowed imports from the Orient. On the journey, Aztec slaves drove mule trains burdened with muslin and silk, crates of jewelry, porcelain, aromatic oils, jade, and ivory. Here they swim the Balsas River, pushing laden rafts before them.

Spanish galleons from Manila unloaded their Oriental wares in Acapulco, where merchants bartered and sold at a trading fair. Colonists bought much of the cargo, but some found its way to Europe via Veracruz and the Spanish fleets. Trade from Spain followed the reverse route to Far East countries.

The merchants themselves risked death in Acapulco from epidemics bred of filthy conditions. One Spaniard called the port an "abbreviated inferno.... All the treasures of this world could not compensate for the necessity of living there or of traveling the road between Acapulco and Mexico."

PAINTING BY JIM BUTCHER

141

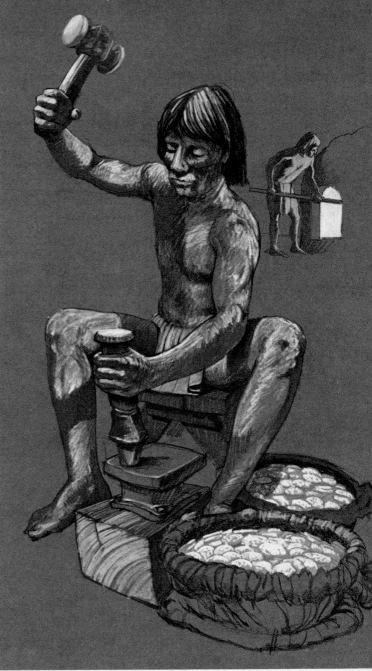

Striking the die with strong, careful blows, an Indian imprints silver coins in Spain's first New World mint in Mexico City. Behind him, another worker melts bullion that he will cast in thin bars, then hammer into strips. Founded by Spain's Queen Johanna, the mint issued its first money in 1536. The earliest coins—later restruck in Spain—had irregular shapes, like the two undated silver pieces at bottom. The Pillars of Hercules mark a two-real piece of the late 1500's (center). In the 1670's the mint began striking gold coins; the doubloon (at top, both sides shown) bears the date 1714.

We found the *Warwick* site in 1966 with a borrowed magnetometer, one greatly refined from the instrument we had used a decade before. Painted yellow and mounted in a fiberglass box on the bow of Teddy's new salvage boat *Brigadier*, the long metal tube looked like a rocket. We named it "The Great Sniffer," painted a face on its mount box, and labeled the marks on the dial "silver," "gold," "brass," and "iron." Soon rumors flew about the islands that Tucker and those Smithsonian people had a new detector that enabled them to clean everything valuable out of the reefs.

But it was good only for iron, and the motion of the boat made interpretation of its signals difficult. Even so, we found several 19th-century wrecks.

Then one day in Castle Harbour, beneath a high coral cliff, the Great Sniffer spoke to us. Quick dives revealed almost nothing. But the next year, a proton magnetometer registered masses of iron on the site. Two days of pumping brought up timber fragments, and we determined that half of a hull lay beneath us.

By measuring that half we reconstructed on paper all of the vessel. The *Warwick*, as she proved to be, still awaits funds for a comprehensive excavation.

Usually wrecks in the West have been severely damaged—by waves and turbulence in shallow, reef-filled waters, or by salvors who have ripped them apart to get at treasure. Frequently storms scatter timbers and cargo over a wide area. Only when my colleagues and I have found substantial remains have we spent the time and money—never inconsiderable—to record the details of a ship and the exact location of artifacts.

In the Mediterranean, archeologists usually find a different situation. Often wrecks have settled to the bottom in deep, quiet water, then moved little or not at all during as much as 4,000 years. Since these wrecks furnish the only substantial data we have on ship construction and nautical practices of classical times, it is important to dig precisely, just as archeologists do on a land site.

One of the oldest shipwreck sites ever found in the New World dates from about 1550. I learned of the site—off Highborne Cay in the Exumas, a chain of islands 40 miles southeast of Nassau—when sport diver Robert Wilke contacted me at the Smithsonian. While spearfishing, he and his friends had noticed unnatural shapes on the bottom, dived down, and found encrusted cannon and an anchor.

A team worked the site for five months in 1967 and brought up ship fittings, two lombards, and a battery of long swivel guns for action at close quarters. We found no other artifacts. The ship probably began sinking at anchor and was cleaned out before going down. We measured traces of timber, and sketched a sharp-prowed, 200-ton ship built light for speed. A striking example of a pirate or privateer ship, I think.

Important archeological material had been found in Florida waters while I was working elsewhere in the 1960's. Kip Wagner had brought up not only spectacular treasure, but also silver plate, navigation instruments, and other artifacts. Most important, from a large lump of clay on the sea bottom, he had dug out three perfect bowls and four cups. All were Chinese, from the K'ang-hsi period (1662-1722)—the first intact porcelain ever recovered from an underwater site, and the first evidence that porcelain was shipped in clay packing to protect it.

Archeological treasures from the waters of Yucatán reefs and Texas sandbars were also brought up in the late 1950's and 1960's. Robert Marx, aided by Pablo Bush Romero's Club of Exploration and Water Sports of Mexico (CEDAM), excavated a trove of European merchandise from a Spanish wreck he had found off Yucatán. CEDAM divers later

Wind-driven waves swamp the Civil War
ironclad *Monitor* as cutters from the steamer
Rhode Island rescue some of her crew. Heading
under tow for Charleston, South Carolina, to
take part in an assault on Confederate
fortifications, the *Monitor* sank off Cape
Hatteras on December 31, 1862. Nine months
earlier she had battled the Confederate ironclad *Merrimack*.

In the summer of 1973 archeologist Gordon
Watts and oceanographer John Newton located
the wreck with sonar. Their underwater photographs show the revolving turret broken off
and partially covered by the overturned hull.
No divers have reached the ship, swept by
strong currents in 220 feet of water.

Lean and swift, the Confederate blockade-
runner *Robert E. Lee* (above) anchors in
Bermuda to take on supplies in 1863. Union
forces later captured the *Lee* and destroyed 85
other such vessels as they approached southern
harbors. Today, divers comb their rusting hulks
for arms, ammunition, and other artifacts.

144

excavated an early 16th-century caravel also sunk off Yucatán. In Texas waters, off Padre Island, a group of businessmen from Gary, Indiana, found and salvaged a treasure ship of the 1553 fleet.

That fleet had followed the usual route from Veracruz to Havana, a half-circle swing along the Gulf of Mexico shores to avoid head winds blowing westward from Cuba. When a storm struck the fleet, the clumsy ships ran aground one after another on the long sandbars lining the Texas coast. The hulls disintegrated and were buried by sand and sea plants, but gold and silver coins regularly washed up on the beaches.

In 1965, fascinated by coins they found at Padre Island, the Indiana businessmen—led by Max and Paul Znika—organized a company to search for treasure. With a proton magnetometer they discovered a large pile of ballast. They used a steel plate to deflect a gentle backwash over a broad area which blew away thin layers of sand and exposed the artifacts without damaging them. Deep sand that shifted with each storm made diving conditions difficult but the salvors recovered hundreds of silver coins, a gold bar, and several silver ingots.

While the treasure was impressive, the artifacts were historically the most important collection from the mid-16th century yet found in the sea—a gold crucifix, crossbows, tools, the complete battery of iron cannon, ship fittings, and above all, three remarkably well-preserved astrolabes, one clearly dated 1549.

Immediately the finders became involved in a struggle with the State of Texas, which claimed all they had brought up. Nearly five

Bricks for ballast, then for building, reach shore 350 years late. Archeologists from an Australian museum recovered them from the Dutch East Indiaman *Batavia*, bound for Java with 300 passengers when she struck a reef and broke up on July 4, 1629.

years later, a court awarded the group $130,000, much less than it had spent on salvage and lawsuits.

The state archeological laboratory holds the collection and is utilizing the latest preservation techniques to care for it.

In the Bahamas, another collection of treasure and historical artifacts waited months in storage until misunderstandings between the salvors and the state could be reconciled. In 1972 Robert Marx, who had worked a number of wrecks, and oceanographer Willard Bascom organized Seafinders, Inc., with the backing of nine investors. After months of hunting, guided by Marx's research, the men found the remains of their target, the Spanish galleon *Maravillas,* lost in the Bahamas in 1656. Company divers began excavating treasure during the summer of 1972. Disagreements over salvaging procedures and custody of the valuables brought a suspension of the lease. Finally, Jack Kelley of Tulsa, Oklahoma, the new company president, negotiated a settlement.

In April 1974, I was called upon to appraise and help divide the collection. During several visits to Nassau, I sat in the vaults of the Royal Bank of Canada and examined a ton or so of treasure: silver bars weighing up to 90 pounds; gold and silver coins, some with unique markings. Among the artifacts were many silver utensils and a variety of ship fittings. With another season or two of excavating, the *Maravillas* will be recognized, I think, as the most valuable wreck ever found.

Archeologists and anthropologists deplore what they consider the ravaging of historic sites by treasure hunters. But the fact remains that until these men began to locate shipwrecks in the 1950's, no one knew exactly where they were, and no state protected its offshore waters as historic areas. Florida passed its underwater antiquities act and hired an underwater archeologist in the mid-1960's, as a result of Kip Wag-ner's find; North Carolina followed in 1967 and Texas in 1972. Alan Albright is developing a program in South Carolina.

All these states have growing undersea programs. Through colleges and universities, the state archeologists organize summer underwater courses for students, but no degree is offered. Only the University of Haifa, Israel, presently has a doctoral program in underwater archeology.

North Carolina's outstanding state program has been going since the early 1960's, when efforts were made to find and preserve the remains of wrecks around the Outer Banks. Among a thousand or so sunken vessels are dozens of Civil War blockade-runners. The state and Duke University sponsored a search off Cape Hatteras for the ironclad *Monitor,* and in 1973 found it.

All of this advances marine archeology, but the problem remains of controlling divers on historically important sites.

I would propose a system of grading wreck sites, and setting aside those that are archeologically important until qualified archeologists using experienced divers can work them. Wrecks of lesser importance could remain open to supervised treasure hunters. Impartial appraisal of the material they recover and a fair division in a reasonable amount of time, a training program in archeology for licensed treasure hunters, and a budget for enforcing the law would gain the cooperation of divers, I think, without destroying their incentive.

Today a dozen qualified marine archeologists are working in the Western Hemisphere, but not a single good museum of underwater history exists. In the next generation, better detecting, diving, and recovery techniques will open up to divers a vast area of ocean in the New World alone. With wise laws and interested educators and philanthropists, treasures that divers bring up can result in a better understanding of our past and an enrichment of our present.

J. N. GREEN (BELOW) AND PAT BAKER, BOTH WESTERN AUSTRALIAN MUSEUM

As rudderfish drift by, a diver cuts the *Batavia*'s massive timbers into manageable lengths, using a chain saw powered by compressed air. After recovery and reconstruction, the ship will become part of an exhibit in the Western Australian Museum in Perth. The brass spigot (far left) came from the *Batavia*, and the candle snuffer from the *Vergulde Draeck*, another Indiaman lost in April 1656. Both ships— with their cargoes and chests of coins—lie in shallow water; the *Vergulde Draeck* sank 50 miles north of Perth, the *Batavia* 250 miles farther up the coast in the Abrolhos Archipelago, a stretch of reefs and barren islets. Discovery of the Dutch wrecks, and the blasting of the *Vergulde Draeck* by treasure seekers, prompted laws protecting historic sunken ships and their relics in waters of Western Australia.

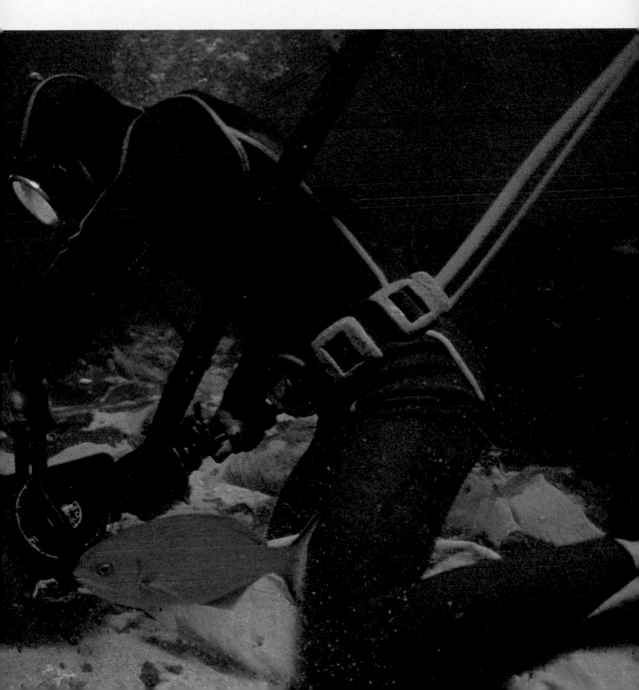

Nature's Jewels Beneath the Sea

7

By Sue Sweeney Abbott

GLIMMERS of pale light from 170 feet above flickered dreamily across the gray wall. Like a tiny oval dish, an orange shell lay face down on a narrow shelf—a glint of gold among the dull-hued sponges. I picked it up. It was only one side of a scallop's fragile shell; still, it was unlike any I had ever seen.

Back on shore, on Andros Island in the Bahamas, I showed it to Tucker Abbott, curator of mollusks at the Delaware Museum of Natural History in Wilmington. "What is it?" he asked.

Feathery black coral, used in making jewelry, rewards a skin-diver off Baja California. Collectors pluck from the sea's bed a variety of rare and exquisite natural treasures.

"You're the shell expert," I said with a laugh. "You tell me."

Surprisingly, he had never seen anything like it among the hundreds of thousands of shells he had studied. Was my find a new species, a treasure of treasures? I hoped so.

But later I learned, a bit sadly, that a collector in Wilmington had four specimens that were found some years before in Cuban waters. My scallop already had a name: *Chlamys multisquamata,* meaning "many ridges." It was so rare, however, that no malacologist had ever seen the living mollusk.

Just two years later, during a diving holiday in the Caribbean, I found and brought up the first—actually the first two—live multisquamatas.

It was as though they had settled down side by side in a coral crevice to wait for me 50 feet underwater near Bonaire, an island off the northern coast of Venezuela. Several divers had passed without noticing them. But after years of watching underwater for colors, textures, and shapes different from their surroundings, I was shell-conscious, and I spotted them.

One appeared to be yawning. In fact it was ingesting water and straining out algae for food. From its bright orange body two rows of eyes the size of pinheads stared out. My own eyes widened in disbelief, for instead of the proper blue eyes of all other known scallops, multisquamata's were brown! I silently named the creatures for myself: Brown-Eyed Susans.

More Brown-Eyed Susans seemed to search me out when I dived in other Caribbean waters. Off Belize, the former British Honduras, my son Wilson picked up a purple piece of plate coral and put it in my collection bag. When I took it out aboard our boat, an orange multisquamata dropped off. Despite its fragility it didn't break, and is my most perfect specimen.

A few other divers have now found them, too, but not enough yet to affect the retail price: $60 for a delicate one-half ounce shell. They're nearly worth their weight in gold.

Only a few of the 50,000 species of marine mollusks are as valuable as that. Of all the world's creatures, only insects appear in greater numbers and varieties than the mollusks. They live in nearly every part of every ocean, from shallow reefs to the sea's deepest trenches.

Scientists theorize that the shell-bearing Mollusca evolved more than 600 million years ago from wormlike creatures that crawled along the ocean floor, feeding on microscopic plankton. When continents formed, and erosion began depositing salts and other chemicals in the sea, the evolving mollusks used them to build shells. Some of the results are spectacularly beautiful.

I have never collected shells for money, but I have met a few divers who supplemented their incomes by picking up rare shells from the ocean floor. In 1971 I was a member of a shell-collecting expedition in the Solomon Islands for the Delaware Museum. Two Australian divers, Wally Gibbins and Brian Bailey, showed me an array of shells they had found. Some of them—Glory-of-the-Seas cones—were each worth as much as the ship propellers the divers were salvaging from sunken World War II vessels.

But the two men had brought up so many *Conus gloriamaris* that they had driven the price down from about $1,000 apiece to $400. Collectors covet this treasure, with its closely packed zigzag of reddish lines on a gold or pale blue shell. Wally told me he had found a colony of them off Guadalcanal, but like anyone who finds a treasure he was vague about the exact spot.

To my astonishment he brought out still more treasured shells—not one but four *Strombus listeri.*

"They're from the Gulf of Thailand," he said. "I traded some of the cones I found in the Solomons for them."

CHARLES PETERSON, MAUI DIVERS OF HAWAII, LTD. (ABOVE) AND RICHARD W. GRIGG

Coral harvesters prepare a bulbous, two-man submersible for launching from its undersea platform. The sub ranges Hawaiian waters as deep as 1,250 feet in search of deep-growing corals such as the pink *Corallium secundum* (top right), rarer and more valuable than black coral. As a blade on the front of the sub snips off a pink coral "tree" at its base, a mechanical claw (right) packs it in a storage basket. After reaping as much as 150 pounds, the craft returns to its platform near the surface, where support divers relay the cache—worth perhaps $4,500—topside. Conservationists hope such collection methods will end the destructive and wasteful system of dredging with tangle nets.

RICHARD W. GRIGG

I told him I'd seen a bidder at a Florida auction pay $1,000 for a listeri a few months before. "Well, they aren't worth that much now. Some fishermen recently trawled up about 500 of them. You can have two of mine for $12 apiece." So it goes in the rare shell market.

Our five-inch, cut-rate treasures each looked like two miniature cones joined mouth to mouth with a wide flare of creamy pearl, like melted ice cream running down one side. Until 1963, when one was picked up off Thailand, only half a dozen specimens existed in collections, and no find had been reported for 80 years.

Our expedition shipped home six chests filled with 400 pounds of shells—600 species—collected during four weeks in the Solomons' hazy waters. We were always careful not to overharvest any area. Some malacologists fear that shell collectors may threaten some rarely-seen species with extinction. But most others consider the numbers that collectors take to be insignificant. The real menace, pollution, kills mollusks by the million.

Only scientists would consider many of our Solomon shells interesting. One, for example, was a drab brown *Vermetus* snail eight inches in diameter. We pried four of the wormlike animals' flat, round shells off the wings of a sunken World War II airplane, where they had glued themselves with surprising strength.

These stationary creatures send out yard-long threads of mucus to which algae stick like flies to flypaper. When the strands are covered with food, the snail draws them slowly back into its mouth. The specimens we collected and preserved in alcohol are now available for study by malacologists from other museums, few of which have even the empty shell.

While in the Solomons I had hoped to spot the extremely rare golden cowrie, a shell owned by fewer than a thousand collectors. But as usual, the elusive *Cypraea aurantium* stayed hidden, so I carried my quest to Fiji, where the island's chiefs once wore the glossy orange shell as a pendant badge of office. I had heard it said that the golden cowrie shell sometimes washes up on the reefs at low tide.

But the Fijians had bad news for me. I would have to dive at night in an underwater cave where the golden cowries hide beneath rocks by day, emerging to feed on vegetation when the cave is darkest. With no divers available to go with me, and with a dangerous current flowing through the cave, I prudently decided not to pursue the shell.

Instead, I found excitement of another sort. On a dive two hours out of Suva harbor I brought up a dangerous *Conus geographus*—the geography cone—deadliest of all the venomous sea snails.

During the boat trip to the site, I had shown a 20-year-old Fijian the mysteries of scuba. He learned so quickly and was so enthusiastic that he went along with me on the first dive, in 30 feet of water.

Suddenly I saw the five-inch cone on the bottom, looking like a slender top on its side. Its shell design suggests white speckled continents and reddish-brown seas. I had seen several geography cones in museum cases. One of these, before it became part of a shell collection, had killed a man with its poisonous, harpoon-like tooth.

Work of Japanese art, these figures emerged from a single piece of red coral in the skilled hands of a carver at the Imai Coral Factory in Tokyo. Selling price: $6,000. At left, Mrs. Masao Imai displays the treasure outside her husband's factory. Archeologists have found ornaments made from red coral at Stone Age sites 25,000 years old. Today, exploitation of the slow-growing creature threatens both the animal and the industry.

From a few feet away I watched, fascinated, as it inched along, burrowing its broad foot into the sand. Two tentacles waved from the tip of the shell, each with a dark eye at the end watching for sea worms or small fish; the creature can harpoon and swallow a fish as large as itself, drawing it inside and digesting it inch by inch. I always marvel that creatures so soft and bizarre mold such beautiful shell houses.

Their secret is the voluminous mantle, a fleshy muscle edged with the open ends of minute tubes. Through the tubes the creatures secrete a limy substance that, spread layer by layer, forms the shell. A final layer gives it its smooth, glossy finish. Other cells can secrete lime to thicken the shell, and even repair broken areas. How the color is applied to make the intricate designs has yet to be explained fully.

My novice scuba companion jolted me out of my absorption, swimming toward me with what appeared to be two feet of rope dangling from his hand. Shocked, I waved my arms violently and gurgled noises through my regulator. I made signs that I hoped meant "get rid of that thing," for I recognized the dangling "rope" as a sea snake, a creature so poisonous that its victim dies in minutes. For a moment I had a feeling of unreality. What were we land creatures doing here at the bottom of the sea, home of deadly reptiles and snails?

My companion got my frantic message and quickly loosened his hold on the snake's neck; it swam docilely toward the surface for a breath of air. I watched until it was well out of sight before cautiously turning back to my own poisonous creature.

Rare coral, a light pink variety known as "Hawaiian angel skin," brings more than $300 a pound—without the gold settings. Often the centerpiece of brooches and pins, this stone supports Hawaii's $8,000,000-a-year industry in coral jewelry.

With my knife I tapped the cone on its large, closed end. Tentacles and foot slid inside. Carefully I picked it up and dropped it into my plastic collection bag. All the way to the boat, I held the bag well away from me and shook it sharply to keep the animal inside its shell.

On board, we gingerly rolled the creature into a can, closed the lid, and breathed a sigh of relief—and triumph.

Out there on the blue Fijian waters I felt that a long time had passed since a day in Bermuda two years before when I had taken my first look at the fantasies beneath the sea. A poor swimmer and frightened of deep water, I had worn a styrofoam belt to support me as I floated face down, looking through a mask.

In the clear water below, sea fans and soft coral bushes swayed rhythmically with the motion of the same waves that gently rocked me. Fish flitted through the submarine forest like bits of rainbow. Enthralled, I scanned the sea floor, hoping to see a shell to add to my collection, begun during vacations on Florida's beaches.

A clam wriggled up through the sand, and I reached for it, forgetting that I was tied to the surface and that the clam was out of reach. An intense longing took sudden hold of me: I must go down there among the sea's creatures.

My family thought me a little mad, or at least obsessed, when I began spending a couple of hours a day at the YMCA making myself learn to swim, and then to dive with scuba. But from my first descent into the amazing seascapes of the Caribbean, it was as if I had begun a new life. Time raced by as I studied the shapes, habits, and habitats of living mollusks, not just their empty houses on a littered beach.

Even in the days when the beach was my only collecting ground, I soon discovered that one living bivalve afforded far more fascination than a bucketful of half shells. A 75-year-old collector, Ted Foley, introduced me to hunting the "live ones" on Sanibel Island, Florida.

I had made a special trip to Sanibel from my home in Cookeville, Tennessee, to enlarge my collection. One day I saw a white-haired man on his front porch sorting an array of king's crowns and angel wings. "Where did you get them?" I asked, expecting the same noncommittal reply I'd had from other shell collectors. But Ted Foley said, "Tomorrow you come with me and I'll show you where they live."

He took me out onto the mud flats at low tide. "Look for a hole about the size of a half dollar," he told me. "That's the opening of the angel wing's siphon. When the tide comes in, water fills the tube and the buried mollusk feeds on algae it strains from the sea."

For a couple of hours we squished through muck, crawled, knelt, sat, and dug. I extracted two six-inch angel wings, their siphons dangling as long as their shells. We washed and opened one of the fragile white shells. With its two sides hinged near rounded tops, it looked like a pair of slender, tapering wings ready for pinning to the shoulders of a cherub.

We picked up king's crown snails, too. Most mollusks come out of the mud to feed at night; perhaps the ones we found were too hungry to wait for sundown. I took one of the largest, a four-inch-long brown conch with gold-tan vertical stripes and a crown of spikes around its larger end.

Ted took me dredging another day, hoping to find a brown polka-dotted *Scaphella junonia*, his specialty. He was slowly putting together a series of these volutes showing the stages of the species' growth, from tiny juveniles to four-inch-long adults. "But the first one we find on this trip will be yours," he promised as he attached a dredge behind his boat. We dragged it for

Black knight of a blue realm, diver Peter Hopper returns
from a successful quest for black coral in 200-foot-deep
water off Andros Island in the Bahamas. He carries a
bangstick for defense against sharks. Largely unexploited
until the advent of scuba, black coral generally lives
below 100 feet in tropical and subtropical waters. In
1972, researchers from the University of Hawaii (below)
found it abundant but slow-growing: only an inch or so
a year. Living black coral appears reddish or dark
brown; washing and drying in the sun reveals the ebony,
skeletal trunk. Jewelers' wheels and buffers transform
the cured coral into lustrous gems. Once thought to
possess curative powers, black coral still goes by the
family name *Antipathidae*—"against disease." Rulers of
some Indian Ocean countries once prized black coral
as the regal raw material for their scepters.

hours along the bottom, but it swept up no junonia volutes. Even after several years I have yet to find a colony and capture one, despite going down with scuba to search the bottom for myself.

On one of my very first dives, however, I fulfilled another longing, one I'd had since I began collecting shells on Florida beaches 12 years before: I brought home a spiny oyster, often called the "chrysanthemum shell" for its long, gently curling spikes, sometimes white, sometimes orange, pink, or lavender. The first *Spondylus americanus* I ever saw had a price tag of $75 in a shop on Sanibel Island.

But in 100 feet of water off Big Pine Key, Florida, a diving buddy and I found hundreds of them growing on mounds of dead coral, and pried loose a few. It didn't matter that sharp-toothed moray eels eyed my probing knife from coral crevices only inches away. I wanted to believe they wouldn't bother us if we didn't bother them, and happily it was true. Nor did it matter that I spent several days cleaning off algae, calcium deposits, and sea plants that almost enveloped the shells.

During my first dives, a hundred feet seemed very, very deep. But soon I was cautiously trying greater depths, down to where the black coral grows.

I remember my amazement at seeing this animal being pulled out of the sea into a boat near the island of Cozumel, Mexico. "What in heaven's name is that thing?" I asked a diver who was dragging up a tree taller than himself with feathery branches a yard long. "Black coral. Found it at 200 feet, growing on dead coral. With a little prying

Underwater prospectors mine northern California's Yuba River, exploring crevices inaccessible to the Forty-Niners. While men at the surface dredge the riverbed for fine gold (top), divers go in search of nuggets. Finds include the 90-ounce chunk at right.

COLES PHINIZY (ABOVE) AND DICK ANDERSON

and a strong pull it came off fairly easily. Some job hauling it all the way up, though," he added. "I'll take it home and make jewelry for my wife." I could see the animal's hard black skeleton where the rusty-red coral polyps had been knocked off the trunk.

Later, I saw small stalks of black coral growing upside down from the roof of an underwater Cozumel cave, then some short bushes off Belize, and a few small trees in the Bahamas. All had rusty-red coral polyps, but in the filtered light below 10 or 20 feet, red looks dark green or black. Some divers reported seeing a forest of black coral at Bonaire island. In its 180 known species, the coral's polyps range from whitish to yellow or reddish-brown.

Black coral grows in tropical and subtropical regions around the world, but mostly in deep water—dangerous for divers with scuba, deadly for free divers. In the Caribbean, divers willing to descend to 200 feet take pieces as souvenirs. Belize is the only country I have visited where the taking of black coral is illegal, but in some other places authorities confiscate it.

But in Hawaii, where in 1958 scuba divers off Maui discovered a bed of black coral —the largest specimens found anywhere in the world—marketing it has become a major industry. Until scuba became available, a piece of black coral cost more than jade. Its family name, *Antipathidae,* reflects the ancient belief that it has the power to charm away all ills.

Now, black coral costume jewelry and small art objects sell for as little as $10, particularly in Hawaii and Japan. Hawaii's merchants informally but effectively regulate the quantity divers retrieve by refusing coral less than an inch in diameter. Prices run much higher for jewelry and art objects made from other, more colorful coral. Japan alone annually retails more than 300 million dollars' worth of art objects made from pink "angel skin" and red coral.

In Wilmington, Joe Felsinger makes earrings, pendants, tie tacks, and other decorative objects from black coral—then gives them away to friends. "They'd be too expensive to sell," he says.

"Black coral is so hard it takes me hours of tedious cutting, grinding, shaping, and polishing to finish a piece," he told me one morning in his lapidary shop.

"I've done it as a hobby since I began diving and bringing up black coral years ago." Joe doesn't dive anymore, but friends keep bringing him coral. "First I put it in bleach to get all the dead animal matter off and kill the smell. Then I let it dry for a year before I cut it. A diamond saw blade is best." Joe slices the trunks into disks, and fills their semicircular cracks with a white pigment and epoxy. "When I'm able to get pearls, mostly from Margarita Island, off Venezuela, I like to set one of them in a disk," he said.

"These days the islanders find only pearls about the size of BB's. The last time I went to Margarita, in the late 1960's, I found them readily available, but just to find one, somebody had to shuck a thousand oysters. Who wants to do that for a pearl market with prices set by the cheaper cultured pearls from Japan?"

In colonial days the Spaniards in the Caribbean forced slaves to dive for pearl oysters and shuck barrelsful, picking out the jewels their masters would take away to Spain. But today few divers go for natural pearls in the Caribbean, or in other once-famous beds like those in the Persian Gulf, the Indian Ocean, and the Gulf of California.

Even in the pearling grounds of the Pacific Islands, the oyster shell—not the infrequent gem—is what lures divers to risk repeated free dives in deep waters.

From March through May in some of the lagoons of the Tuamotu Archipelago east of Tahiti, hundreds of Polynesians each bring up 50 or so of the large, flattish shells during a day's work. But they have to make about 40 dives to do it; scuba is not allowed, and only shells more than five inches long can be taken. Merchants stake the divers, and sell the mother-of-pearl in Tahiti for a worldwide market. Since 1965 the number of shells taken has dropped sharply: Young men prefer city jobs to diving.

Someday, I would like to go to Takaroa Atoll in the Tuamotus and watch from underwater as goggled divers, holding their breath and clutching weights to speed their descent, fall rapidly 120 feet to the bottom of the lagoon, twist loose one or two eight-inch shells, and race hand-over-hand for the surface. In those clear, warm depths I might also see a natural treasure as spectacular as one I came across during a dive in a very different sea: the cold, turbid waters off the New Jersey coast.

On a sunny day in April 1974, I was swimming among the tangled metal scraps of a U. S. merchant ship sunk by German torpedoes in 70 feet of water during World War II. Like most of my dozen or so diving club companions, I was shining my flashlight through the green water—made murky by clouds of silt and plankton—searching for lobsters that hide in crevices formed by the twisted metal. We look for their shiny eyes or their slender antennae, then grab them firmly behind the head. Big ones resist with great strength, their crusher claws snapping wickedly. I've seen 20-pound giants that measure a yard from claw to tail tip.

As I approached a jumble of metal, its encrusted surfaces resembling rough concrete, I played my light slowly back and

Once a rare prize for shell collectors, the precious wentletrap—Dutch for spiral staircase—plummeted in value when new shell beds turned up in the 19th century. According to legend, Chinese artisans once fashioned counterfeit wentletraps from rice-flour paste; such fakes would now command far higher prices than the genuine shells ever did.

EPITONIUM SCALARE, APPROXIMATELY LIFE-SIZE

forth. Suddenly, through the murk, the beam hit a ghostly head of lacy hair, tangled by the movements of the water.

The lace, I saw, was an oval mass of feathery, curling, four-inch-long tentacles: a white sea anemone the size of a dinner plate growing on the wreckage. In warm, clear seas I had often seen smaller anemones of various colors, but never imagined that I'd find one so large in such frigid waters.

I stroked the tentacles with a gloved hand — some anemone tentacles can sting bare skin. They withdrew a little inside the thick, columnar body. As I swam away I wondered how long this simple polyp had lived here, its eerie beauty unseen. It might have been older than I, judging from its size and the fact that sea anemones in aquariums have lived 70 years.

Nature's undersea treasures often mystify me. I remember vividly an incident during a dive in the Ionian Sea. Swimming lazily near fragments of clay pots off the Greek island of Cephalonia, birthplace of Ulysses, I wonder briefly if they were part of an ancient cargo. Finished with my checkout dive for some archeological work I will soon be doing, I indulge myself in a brief search for sea shells. The sun's rays trickle down through 30 feet of clouded blue water to nearly barren sands. The hunting will be poor. I see just one or two small *Murex* shells, source of a dye used thousands of years ago for cloth worn by the rich and powerful; its color gave rise to the phrase "to the royal purple born." As early as 1600 B.C., the dye was used in Crete. Later the Phoenicians monopolized its manufacture. They cracked the shells to extract the mantle; in sunlight its mucus changed color from cream to purple-red. After three days in saltwater, the body was then boiled in the water. Wool dipped into the fluid took on deeper and deeper hues of brilliant magenta.

In the distance I see the light reflecting on a slender, fan-shaped object. Seaweed?

A piece of metal? I swim closer. My heart, already slowed 20 beats a minute by the water's pressure, seems to skip a beat entirely. Regally straight and wafer thin, a foot tall and some eight inches wide at its top, it is a large *Pinna nobilis,* a pen shell found only in the Mediterranean.

I begin to dig, making the sand fly in gritty clouds. Baring another 12 inches of the tapering shell, I work loose its stringy bronze-colored roots, as long as my hand. Back on the ship, I pry open the shell and remove the skimpy, oysterlike body of the mollusk.

The root fibers of the shell were the raw material — according to some historians — of Jason's fabled Golden Fleece.

In ancient times, pen shell fiber was washed, dried in the shade, combed and carded, then knitted or woven into clothing destined for kings. Strong and resilient, the threads were made into silky gloves or stockings, lustrous bronze-gold collars, and caps for royal heads.

Sometimes I give the treasures I find to the museum, where they are available for study or for display. Occasionally I make a small contribution to malacology by finding a shell in a previously unknown habitat. Adding my shells, eyewitness accounts, and photographs to scientific knowledge gives value to my collecting — while collecting gives purpose to my diving.

But my most treasured collection is in my mind — the remembered images of sea creatures I have merely observed. The true reward for me is to glide about whenever I can in a liquid world of enveloping beauty.

Often when I dream, I am swimming with the fishes through astonishing forests where the trees are animals, where rocks are growing sculptures coated with flower-like creatures, where sea shells are gaudy mobile homes built by flabby architects. And when I slip below the surface of the water, earthbound cares vanish as I pursue the game of discovery in a realm of ceaseless wonders.

Covered by marine growth, the limestone wall of Andros Island in
the Bahamas plunges into darkness.
Breathing a mixture of helium and
oxygen through electrolungs, divers can
follow it down as far as 400 feet. Near
that depth in 1971 divers found the rare
sea snail below, relative of a generally
extinct group of snails some 200 million
years old. Collectors will pay $500
for a specimen.

Collecting shells as a hobby began long
ago. One early enthusiast: the mad Roman
Emperor Caligula. Advancing on Britain
in A.D. 40, he stopped at the English
Channel, set his legions to gathering
shells, and returned to Rome with "the
spoils of conquered ocean!"

Elizabeth Bligh, wife of the *Bounty*'s
captain, assembled a large collection,
partly from shells brought home from
the Pacific by her husband.

WALTER A. STARCK II   PLEUROTOMARIA ADANSONIANA
APPROXIMATE DIAMETER 4 INCHES

PAUL A. ZAHL, NATIONAL GEOGRAPHIC STAFF. CYMBIOLACCA PULCHRA, APPROXIMATE LENGTH 2 3/4 INCHES

CALPURNUS VERRUCOSUS, APPROXIMATE LENGTH 1 1/8 INCHES

MUREX TROSCHELI, APPROXIMATE LENGTH 3 1/2 INCHES

Movable feast of colors glides across the ocean floor: sea snails stalking their prey. These four gaudy carnivores belong to a family that numbers in the thousands of species; a few rarities bring hundreds of dollars apiece from collectors. A Halloween-hued volute (left) creeps along exposed rock. The false cowrie (bottom left) wears polka dots on both its shell and its fleshy mantle. The spiny-backed *Murex* (middle) yields juices that ancient Phoenicians, Greeks, and Romans used for making purple dyes. Whorled, deep-dwelling sundials (below) live only in the tropics; collectors covet some of the rare species.

Overleaf: Splash of scarlet beneath a spotted shell marks a sieve cowrie. Tentacles extended, it forages for marine creatures on a colony of coral in the Philippines.

PAUL A. ZAHL, NATIONAL GEOGRAPHIC STAFF (OVERLEAF)
CYPRAEA CRIBRARIA, APPROXIMATE LENGTH 1 1/4 INCHES

ARCHITECTONICA PERSPECTIVA, APPROXIMATE LENGTH 2 1/2 INCHES

TEREBRA SUBULATA,
APPROXIMATE LENGTH 6 INCHES

NATIONAL GEOGRAPHIC PHOTOGRAPHER BATES LITTLEHALES (ABOVE)
AND PAUL A. ZAHL, NATIONAL GEOGRAPHIC STAFF

NAUTILUS POMPILIUS, APPROXIMATE DIAMETER 7 INCHES

CHLAMYS NOBILIS

In the sands of Australia's Great Barrier Reef, malacologist Phil Coleman searches out the compact, functional shells formed by the limy secretions of mollusks. The auger shell (far left) burrows with its sharp tip to escape enemies. Cutaway view of a chambered nautilus (bottom left) shows tubes that permit the animal to regulate the amount of buoyant fluid in each chamber; the nautilus dives as deep as 1,800 feet. A scallop (middle) wears a shell with ridges resembling a patch of red cabbage. Intricately patterned harp shell (bottom right) has glossy, black-striped ribs. The female paper nautilus—or *Argonauta*—fashions a shell (below) that serves as both home and egg case.

ARGONAUTA NODOSA, GREATEST PROPORTION 8 INCHES

HARPA ARTICULARIS, APPROXIMATE DIAMETER 3 INCHES

171

PETER STACKPOLE

Future Challenges, Lost Fortunes

8

BY ROBERT STÉNUIT

ON a deserted beach in western South Africa, silver ducatoons wash ashore this very minute, the legacy of an East Indiaman sunk nearby. No one picks them up because no one lives there; the inhospitable land affords neither water nor food. An expedition to retrieve the treasure would have to import—at great expense—all life's necessities. Furthermore, the beach belongs to a diamond company, whose planes regularly patrol the area, shooting at trespassers. The hardships and risks are great. And so the

Off Bermuda, a diver cleans an artifact from the Confederate blockade-runner *Nola*, wrecked in 1863. Worldwide, marine archeologists work to improve detecting and recovery techniques.

172

ship sits, untouched, quietly spilling its treasure onto the barren sands.

Somewhere on the other side of the Atlantic lies the fabled fleet of Francisco de Bobadilla, onetime governor of Hispaniola. Thirty ships set sail for Spain in 1502, loaded with New World plunder that included a gold nugget big enough to eat from. All but four ships were lost, some in the Mona Passage between Hispaniola and Puerto Rico during a hurricane. Divers could search the passage and find some of the ships, but their job would be so monumental that even Bobadilla's golden nugget might not pay the enormous cost of recovery.

Then there's *Cinque Chagas*, a huge Portuguese carrack. Returning from the East Indies in 1594, she ran up against a pack of English warships off the Azores. Proudly fending off her smaller attackers, she fought for days, finally choosing to sink rather than surrender. Aboard was a truly fabulous bonanza of gems, gold, silks, artwork, and porcelain from the Orient. But the English victors failed to note the exact position of *Cinque Chagas*, and logbooks show she sank in blue water thousands of feet deep.

Such are the treasures that set divers' mouths to watering. Unfortunately, all these ships may be lost forever, for in addition to other difficulties, too little is known of their exact whereabouts. Search the ocean floor? I'd as soon sift the Sahara for a ruby.

Yet divers continue to find virgin wrecks, simply because so many exist. More than 20,000 ships have gone down in this century alone. Of history's hundreds of thousands of shipwrecks, only a few thousand have ever been found.

Why? Quite simply, treasure diving entails incredible expense, dedication, and hard work—both mental and physical. It has to be a full-time job if it is to succeed. And not everyone is prepared to pay the price—to sever all the ties that hold him to the normal world. But if you insist on throwing your life savings into the ocean, begin by choosing a wreck.

The temptation is strong to look for deep, untouched wrecks, but I prefer shallow ones: They are much easier to find and excavate. Partly salvaged hulks are my favorites, not only because they are shallow but also because they have often spawned lawsuits for embezzlement, theft, and other crimes. Court records are a prime source of reliable information on both what the ship carried and what was salvaged. In contrast, deep wrecks usually leave no survivors or witnesses; thus there are no firsthand reports of their whereabouts.

Generally, the rules are the same for all wrecks, deep or shallow. First you study the ship's history, gleaning all you can from national libraries, archives, and museums. Suppose you pick a Spanish galleon to hunt. You then must be proficient in paleography —the reading of old handwritings—and know 16th-century Spanish *before* you begin to pore over the moldy records in perhaps half a dozen different archives. You cannot afford to stop your research until you've found every scrap of information on the ship and its cargo. Long before you seek your wreck in the sea, you must find it on paper and on inaccurate, antique maps with long-forgotten place-names. And no, you cannot hire someone else to do it for you. Others will not go to a tenth of the trouble you should, nor have the stubborn persistence needed to do the job thoroughly.

But let's be optimistic and suppose that your research is behind you; you think you've located a treasure-laden wreck and
(Continued on page 178)

Overleaf: In an underwater classroom, archeology students view grindstones lost in 1818 when the British merchantman *Caesar*, bound for Baltimore, struck a reef near Bermuda. Programs like this train increasing numbers of scientists to explore the world's seas.

PETER STACKPOLE (OVERLEAF)

Atmospheric diving suit, with a self-contained life-support system for the operator inside, emerges from a test tank in a laboratory in Farnborough, England. The suit maintains sea level pressure for the diver at any depth down to 2,000 feet, allowing him to ascend without decompression. It takes its design from the "Iron Man" unveiled in 1930 by British inventor Joseph S. Peress (below). Interest in his concept revived with advances in metallurgy. Made of lightweight magnesium alloy with aluminum alloy joints, the apparatus has handlike manipulators for picking up objects or holding tools. Divers will use it to reach deep-sea wrecks and mineral deposits.

are eager to strap on your scuba. Not yet. It's time to study in detail the local laws governing underwater salvage. What you find undersea is rarely yours to keep, for most nations claim all ownerless wrecks within their waters. Anything brought up from them belongs to the government, not you. You, as excavator, are entitled to a salvage fee—but the fee varies immensely, from zero to one hundred percent of whatever the ship brings at auction, or is estimated to be worth by the state, when it is the buyer. And suppose a Portuguese ship sank in Dutch waters—you've got *two* governments to deal with, and the laws pertaining to wrecks can be as tangled as a kelp bed. Often they're recent or vague or contradict each other. Yours may be a test case, so brace yourself for years of court fights.

Then there is the expense. Equipment and men cost quite a lot for a season of diving. How much is left of your life savings at this stage? But again, let's be positive. You've found a wealthy benefactor, or perhaps you are a millionaire. You've got the money, you know the laws, and you think you know where the wreck is. Time to equip your expedition.

Choose a sensible boat for your search. It's best to fight down the romantic urge to captain a schooner—they're pretty, but expensive and terribly impractical for diving. Nothing beats a large inflatable raft; it can carry a ton of gear, it needs no harbor, it

NATIONAL GEOGRAPHIC PHOTOGRAPHER BATES LITTLEHALES

NATIONAL GEOGRAPHIC PHOTOGRAPHER OTIS IMBODEN

X-rayed flintlock pistol shows clearly through an encrustation of coral. Diver Art McKee examines the film, and finds powder and ball still in the weapon after nearly 2½ centuries in the sea. The gold crucifix, from a wreck of the 1715 Spanish fleet, bears a tag identifying it as property of the State of Florida. Under a law passed in 1965, the state regulates the activities of treasure divers: A state agent must accompany all crews, and Florida takes possession of the treasure while salvage continues.

won't overturn, and it survives repeated beachings and bashings by the surf. Whenever possible, you should sacrifice glamor for basic needs, so don't plan on living aboard. Settle ashore in an unromantic but warm cottage with central heating; you'll need it after days of working in the cold ocean. And if the wreck site is remote—as is often the case—you may have to take along food and other supplies.

What else should you take? Today's well-equipped—and well-financed—diver can choose from such aids as underwater television cameras, suspended diving chambers, even small submarines that serve as observation posts, living quarters, transportation, and decompression chambers all in one. These are useless for locating wrecks, but can be a help in identifying a target, or surveying and recording before excavating.

Tethered manipulators are mobile, unmanned devices with steel arms and mechanical hands. They're controlled by an operator on the surface watching a television monitor. The Navy's Nedar II, for instance, will recover test torpedoes from depths of 20,000 feet.

Such devices have proved very valuable for attaching a line to a lost object—an H-bomb or an incapacitated small submarine. However, they are powerless in front of the mountain of mud, rotting wood, and rusted iron that constitutes a deepwater treasure wreck.

I think only one innovation—liquid breathing—would truly aid the individual treasure diver because, unlike gas breathing, it would allow long working time at the bottom without decompression. At Duke University, Dr. Johannes A. Kylstra is exploring the possibility of flooding the diver's lungs with a hyperoxygenated, slightly saline solution. With equipment to heat the liquid to body temperature, to pump it to

and from the lungs, control the oxygen and carbon dioxide, liquid-breathing man may one day swim and work freely at any depth.

Modern electronic detection devices scan the sea bottom for wrecks and are nearly infallible—that is if you're after a large steel ship conveniently lying on flat sand.

But if you're seeking a galleon, the odds are that your target will be a shapeless, brine-soaked mass of rotting wood buried deep in mud, or 10,000 splinters scattered in rocky crevices. A sonar pinger might detect a waterlogged mass if you know precisely where to "ping," for it has no lateral vision. Metal detectors scan a very narrow range and penetrate only a few feet of sea floor. Magnetometers detect large iron and steel targets reliably only if the bottom is reasonably smooth and magnetically constant. I have worked with six different magnetometers and never found the wrecks I was seeking, only modern ones I didn't want. And the future promises little improvement, unless entirely new systems are invented. Present breakthroughs have their applications, but none is a panacea. Even the most expensive magnetometer, which costs thousands of dollars a day to lease—along with the ship it's installed in and an operator—can fail because the treasure hunter is looking for odd bits of wreckage that don't give recognizable signatures. That's assuming the equipment works at all. It always worked best for me when the inventor himself was there to keep his fingers on the knobs, adjust everything, and change burned-out circuits as many times a day as necessary. Despite all technological advances, the most reliable tools for identifying and excavating a wreck remain the eyes and hands of an experienced diver.

Digging out a wreck is a very slow job, requiring planning and patience. There are no hard and fast rules, for success once does not mean continued success. Techniques differ from site to site, but the best overall

equipment I've found are hand tools—crowbar, hammer, chisel—and simple machines like an airlift to vacuum debris. A high-pressure water jet will flush away the mountains of sifted tailings from a dive; hydraulic jacks can ease out boulders, and inflatable neoprene bags will float cannon to the surface or shift heavy objects during the dig. Very small explosive charges, carefully placed, will safely dislocate large lumps of the natural concretions that artifacts are usually embedded in. Together, they comprise the essentials in a modern treasure diver's toolbox.

Back to our hypothetical expedition. You have completed your research. Your equipment is ready and waiting. The hired divers are eager to go over the side. Time to dive! You are, of course, an expert diver, sailor, and navigator. Nothing scuttles a salvage job faster than inexperience. For months, perhaps the better part of a year, you will work on this one wreck, carefully exposing each section, noting every artifact you find, trying to decipher as much as you can from these disjointed clues dredged from the sea. Until you've sifted all the debris, you'll never know for sure whether you have all the cargo. The sea may have destroyed much of it, certainly the perishable silks, woods, and artworks, unless they're well covered by mud or sand, or embedded in natural concretions. The jewels and gold long ago spilled out of the rotting containers in which they were packed. Has the sea's swell scattered them beyond recovery? Or has some unrecorded looter already

Mammoth bones 50,000 years old attract scientists to the bed of Florida's Aucilla River. Above, paleontologist S. David Webb raises a leg bone attached to an air-filled plastic bag. Dr. Webb calls his state's rivers "a neglected fossil resource. Paleontologist-divers may phenomenally increase our knowledge of extinct animals and of the evolutionary process."

pillaged your wreck? You won't know the answers to these disturbing questions until you actually excavate.

And when you find your wreck, she may of course be the wrong one—or the right one but without treasure, such as *Trinidad Valencera*, one of three Armada ships found since I discovered *Girona*. Her main treasure was unloaded the day before she sank.

The sad truth is that most wrecks—even treasure-laden ones—yield the finder far less net profit than expected. So the diver must always balance his possible find against the staggering costs of the expedition. Even though my patron supports all of my excavations, I make my living from articles and films about diving, not from the unreliable wrecks themselves.

What about all those multimillion-dollar finds in Florida and the Bahamas, you say? Well, hard-won success understandably generates enthusiasm, and the sound of success attracts new investors. All exaggeration apart, you must first subtract the coins that were damaged by erosion and corrosion in the sea. Don't forget, too, that often when a diver says his 25,000 coins are worth a million dollars, that means they would be worth that much if each could be sold for the going price *before* he made his find. But his cache has ruined the market; the more coins recovered, the less rare and valuable each will be. So what can he do but put what is left of his treasure in a museum and charge visitors to see what an imaginary million dollars looks like.

Suppose your treasure auctions off for $100,000—a very good price. You must deduct the auction house commission: 10 to 25 percent. From the amount left you must pay the share of your host government, such as the 25 percent Florida keeps of any treasure found in its waters. Then there are expenses, salaries of the crew and divers, the amortization on your expensive salvage equipment, and perhaps court costs for a few lawsuits contesting ownership of your treasure . . . and your "fortune" shrinks rapidly to a pittance.

Finally, if your wreck is important, I suspect that all those years of chasing rainbows will have attached you too permanently to your pot of gold and jewels and the period of history into which they plunged you. You will not want to sell your finds piecemeal, but instead will settle for whatever a museum can pay, just to keep the treasure together. You do this even though you know you will get less money, making this year's debts and the financing of next year's expedition even more of a problem.

Have no salvage operations yielded solid fortunes? Maybe a few, like Kip Wagner's or Teddy Tucker's, but even they recovered nothing like the 26 tons of silver found in 1687 by William Phips. Galleons salvaged today rarely pay off so well. Some salvors search for modern bullion, such as the five tons of gold and ten tons of silver recovered in the 1930's from the English liner *Egypt,* hit by a freighter in fog off Brittany.

In 1973 a large British salvage firm raised more than half a ton of gold bars from the freighter *Empire Manor* sunk 330 feet down in 1944 some 200 miles off Newfoundland. But such finds are rare. Today's salvage fortunes usually spring from less glamorous cargoes, such as the thousands of tons of tin, copper wire, brass ingots, and lead pigs, worth millions of dollars, that have gone to the bottom during wartime. Such hoards are for professionals to recover, not for you or me. Deep-sea salvage companies rarely use divers when they can rely on observers in suspended chambers to direct the placing of explosives, the dissecting of the shattered hull, and the bringing up of the cargo. They are businessmen, not poets dreaming of long-lost doubloons and ducats.

And what of other treasure ships, the thousands that set sail and never returned? Most went to the bottom along the world's main trade routes and beneath the great battle sites of the sea. Throughout the 16th, 17th, and 18th centuries, hundreds of tons of gold and silver rode the Gulf Stream from Mexico to the coffers of the Spanish kings. This financial lifeline sometimes injected more gold into Europe's economy *annually* than all of Europe possessed at the time the Americas were discovered!

While storms sank numerous treasure ships, the precious cargoes that arrived safely often were reshipped in East Indiamen bound for the lucrative spice trade of the Orient. Thus Neptune had a second chance to grab the Spaniards' silver. Year after year, he battered those ships, turning the world's trade routes into treasure lanes paved with bullion. On them, thousands of galleons and East Indiamen lie camouflaged by silt, sand, and marine life, waiting for the divers of the future.

(Continued on page 188)

Pit-marked isle of mystery in Nova Scotia's Mahone Bay has intrigued treasure hunters for nearly two centuries. In 1795 three boys digging in a hollow beneath a tree on Oak Island found a clay-lined shaft and two floorings of oak logs, one ten feet below the other. Nine years later, two of the three resumed their excavation. Every ten feet they encountered oak-log platforms. Then, at a depth of 98 feet, the pit flooded. They gave up, but others followed. Searchers in the mid-1850's discovered an ingenious system of drains and tunnels that channeled water to the shaft from a cove 520 feet away. Drills brought up a fragment of vellum with writing on it, pieces of tropical fiber, and at 172 feet hit an iron floor. Theories of what may lie hidden in this formidable underground complex range from Inca gold to the booty of the pirate Captain Kidd. Daniel C. Blankenship, in wet suit, has twice dived into the murky shaft. David Tobias, far right, organized a consortium of U. S. and Canadian backers, and has raised $300,000 in an effort to solve the enigma, one of the greatest in treasure lore.

JAMES H. PICKERELL, BLACK STAR

183

JAMES A. SUGAR (OPPOSITE) AND N.G.S. PHOTOGRAPHER OTIS IMBODEN

Remnants of a Bronze Age civilization emerge from thick volcanic ash on the Greek island of Thera. An eruption about 1500 B.C. destroyed part of the island and its Minoan civilization. Frescoes and pottery (above) date the disaster and attest to the artistry of the Minoans. Eleven centuries later the Greek philosopher Plato wrote of the cataclysmic end of Atlantis, an island-centered empire that "was swallowed up by the sea and vanished." Some historians believe Thera inspired Plato's story of the "lost continent."

Overleaf: Near Bimini in the Bahamas, a diver brushes sand from limestone blocks. Early archeologists first thought them evidence of the legendary Atlantis; geologists consider them natural formations—part of a shoreline thousands of years old.

NATIONAL GEOGRAPHIC PHOTOGRAPHER BATES LITTLEHALES (OVERLEAF)

In years to come who will look for them, and for others scattered under the seven seas?

It's safe to predict that Teddy Tucker will make regular finds on Bermuda's reefs, and Bob Marx will keep checking out one after another of the ships on his list of wrecks of the Western Hemisphere. Half a dozen salvors will keep Florida in the news.

Rex Cowan, a London solicitor turned treasure hunter, will have plenty to do excavating two Dutch East Indiamen, the *Hollandia* and the *Princess Maria*, in the Isles of Scilly. Colin Martin, director of Scotland's St. Andrews University Institute of Maritime Archaeology, seems bound for life to the Spanish Armada, with three more ships now being slowly excavated.

All the Armada ships now located are of different types: *Girona,* which I found, was a galeass; *Nuestra Señora de la Rosa,* found off Kerry Head, Ireland, was a war galleon; the *Gran Grifon,* off Fair Isle, northeast of Scotland's Orkney Islands, was a chartered Baltic trader; and the *Trinidad Valencia,* off Inishowen Head, Ireland, was a Venetian merchantman loaded with war supplies, field guns, and siege equipment.

From *Trinidad*'s sandy grave Colin and his associates have already raised superbly decorated brass guns and many other artifacts. They are in pristine condition. Such a massive mine of information will make the scattered remains of *Girona* look very unimportant in years to come.

Replica of the *Golden Hind* sails in tribute to Sir Francis Drake, first Britisher to circle the globe, and his country's most famous privateer. For 30 years he preyed on Spanish ships for his Queen, Elizabeth I. Drake's last voyage ended when he died of dysentery aboard the *Golden Hind* in 1596 near this rock off Portobelo, Panama. Here his crew lowered his lead coffin into the sea, then honored him by scuttling two royal vessels and several captured Spanish galleons. Divers seek the coffin for the artifacts they hope to find inside.

In another part of the world, studious Scandinavian divers will continue to salvage historic wrecks, unconcerned with the market value of the ancient gold and artifacts they bring up. Australians, guided by Jeremy Green, curator of marine archeology at the Western Australian Museum in Perth, will continue to piece together the history of the Dutch East Indies trade from the treasures they find and recover.

And me? My future includes a Dutch East Indiaman that succumbed to the failings of a bad pilot and—though slated for the Orient via Africa—sank in rocky shallows near Madeira in 1724. She was partly salvaged a few years after the sinking, but enough of her cargo of silver bars, coins, and manufactured goods remains to draw me there.

Then there's *La Comete*—The Comet—a French privateer sailed by the swashbuckling Knights of St. John of Malta, who terrified their Turkish enemies on land and sea during the 16th and 17th centuries. Lost off Crete, *La Comete* should yield some gold and silver, but its chief attraction lies in its wealth of artillery, navigational instruments, medals, and other historically valuable objects. If discovered, she would be the first wreck representative of a fascinating period in history, when the Knights preyed on the ships of the Turkish Moslems from their stronghold on Malta.

I'd love to find the *Soleil d'Orient*—Sun of the East—a French East Indiaman that carried an enormous load of riches with the first ambassadors of the King of Siam to France. I have read and reread the bills of lading, dreamed of *Soleil's* hundreds of gold and silver vessels, its many *objets d'art* chosen from the royal Siamese treasury to impress the court of Louis XIV. But where to begin a search for a ship that sank in the 1680's, unseen and with no survivors? We know only its cargo and the day it set sail. The entire ocean is our hunting ground.

Someday I'd like to turn to earlier ships, such as those that ferried Europe's medieval Crusaders across the Mediterranean to the Holy Land. No wreck from this era has ever been found, and very little is known of the armament or weapons of the time. A find would be exciting, for it would fill historic gaps, if anything aboard has survived the centuries.

Just as the sea is full of wrecks, so are its tributaries, the world's major rivers. When King Louis XIV of France coveted some exquisite marble statuary found in an ancient Roman villa, the statues were shipped to Bordeaux, then out the Gironde estuary toward the Seine River—the route to Versailles. The boat sank. No records of its site exist, and today tons of silt probably bury it. One could dredge the entire estuary and still not find the statues, for by now they may lie under the ever-changing bank.

In the mid-1800's, during the reign of Napoleon III, a shipload of priceless Assyrian art treasures, carved sphinxes and lions and such, went down in the Tigris River on its way from Iraq to the Louvre. It was never recovered. The statues were made of a particularly soft stone, known to dissolve rapidly in water. Even if the ship remains, the treasures have melted.

Such is the future of treasure diving. Many wrecks are left, but none is easy. Each has a catch of some kind: It is too deep, its location too fuzzy, its treasure too dubious or unsalable. Many of the shallow, easy-to-find wrecks have been found. But not all of them. Numerous ships sunk in shallow water remain to be discovered.

If you find one, your sense of achievement will be beyond description. But you'll also learn that the joy of treasure diving lies not so much in finding your ship—though that is a pleasure indeed—but in searching for it, in probing the dark green fantasyland for the hidden, encrusted debris that might, just might, yield a fortune in treasure.

GEOGRAPHIC ART

For Treasure Divers
of Tomorrow,
A World of Wrecks

*Distinguished both as
marine archeologist and writer,
Robert Marx has accumulated information
on more than 25,000 sunken wrecks.
For this map,
he selected 20 rich, unsalvaged sites.*

1 California In 1582, after a four-month crossing from the Philippines, the Spanish galleon *Santa Marta* sank in 200 feet of water off the north end of Catalina Island. Her cargo: gold and silver plate, hundreds of chests of Chinese porcelain, silks, and spices, and chests of gems and jewelry.

2 Massachusetts Lost in fog, the pirate ship *Whidah* struck a sandbar half a mile off Wellfleet on Cape Cod in 1717, carrying a fortune in silver, gold, jewels, and ivory.

3 Delaware The warship H.M.S. *De Braak,* bulging with captured Spanish silver, gold, jewels, and 70 tons of copper ingots, entered Delaware Bay for supplies in 1798. A squall capsized the ship as she prepared to anchor, about a mile east of the tip of Cape Henlopen.

4 Bermuda The Spanish merchantman *Santa Maria de Reposa,* her rudder damaged by a hurricane, drifted for weeks in the Gulf Stream before breaking up on a reef five miles west of Somerset Island in 1550. Salvors never found the *Reposa,* or her cargo of gold and silver.

5 Bahamas In 1765 the galleon *Santiago El Grande* set sail from Havana carrying treasure valued by the Spanish at more than four million pesos—until the mid-1800's, *peso* referred to a unit of silver weight: about an ounce. A hurricane drove *El Grande* up the Great Bahama Bank and wrecked her in 17 feet of water seven miles southwest of Beach Cay, north of Riding Rocks.

6 Cuba A storm wrecked five ships of the 1711 Spanish fleet off the port of Mariel, 20 miles west of Havana. One—the galleon *Santísima Trinidad* —carried more than three million pesos in treasure. Salvors recovered about half of it.

7 Hispaniola The 1724 Spanish treasure fleet, sailing for Spain, was struck by a hurricane near Bermuda. Those ships that did not sink headed for Santo Domingo, where some broke up near Samana Bay. The flagship, *Nuestra Señora de Guadalupe y San Antonio,* and tons of treasure sank in 200 feet of water off Cape Samana.

8 Dominica As the 1567 Spanish fleet prepared to sail from Veracruz, Mexico, word came of two English fleets waiting to intercept it. To avoid them, the Spanish hugged the coast of Yucatán, then turned east across the Caribbean. A storm smashed six treasure ships onto the northwest tip of Dominica. Carib Indians killed and ate many of the survivors, then recovered some treasure and buried it in caves. The wrecks—in 100 to 200 feet of water—eluded Spanish salvors.

9 Venezuela In 1815 a large fleet of Spanish warships sailed to Venezuela to suppress a revolution. While anchored off Coche Island, one of the largest—the *San Pedro Alcantara*—caught fire and exploded. The blast scattered more than 800,000 silver coins over a wide area; fewer than half were recovered. Amateur divers from Caracas have recently salvaged a few cannon and artifacts.

10 Colombia The *San José,* richest single Spanish treasure galleon ever lost in the Western Hemisphere, sank in 1708 with silver worth 11 million pesos. While battling an English squadron, *San José's* powder magazine exploded; the ship and 600 crewmen went down in 550 feet of water off Treasure Banks, 17 miles west of Cartagena harbor.

11 Serranilla Bank Struck by a hurricane in 1605, four silver-laden galleons smashed onto the reefs of Serranilla Bank in the western Caribbean. In 1667 Cuban fishermen came across the site and recovered a few silver coins. During the next six years the Spanish tried to salvage the wrecks, but bad weather and pounding breakers stopped them.

12 Ecuador On a rainy night in 1654, the flagship of the Spanish treasure fleet struck a reef at Chanduy Shoals, near the mouth of the Guayaquil River. Within a few days, shifting sands covered the wreck, halting the work of the salvors. Today, storms wash gold and silver coins ashore.

13 Ascension Island Bound for Lisbon from India and the Far East in 1568, the Portuguese carrack *Nossa Senhora da Estrella* broke up off the northwest coast during a storm, carrying treasure valued at four million cruzados—a Portuguese gold cruzado in the 16th and 17th centuries equaled in value about an ounce of silver. Twenty passengers eventually were rescued. Artifacts from *Estrella* occasionally wash up on the beach.

14 Spain Perhaps one of the world's richest treasure sites, the mouth of the Guadalquivir River holds the wrecks of more than 200 ships lost while trying to navigate the shallows. One of the richest: *Nuestra Señora de la Navidad,* with nearly 4½ million pesos in gold and silver.

15 Azores Mariners sailing from the New World and the East Indies used the Azores as a navigational checkpoint; hundreds of ships came to grief on the rocky shores, sinking in waters too deep for salvors of the time. In 1591 a hurricane drove about 50 Spanish ships onto the western end of Terceira Island; 14 carried gold and silver.

16 France Fleeing the turmoil of the French Revolution in 1790, the *Quintadadoine* sailed on the Seine from Rouen with jewels and other treasures belonging to Louis XVI and his nobles, plus the gold and silver of two wealthy abbeys. Approaching the sea and safety, the ship ran aground and broke up near the village of Quillebeuf.

17 Netherlands In 1799 the British frigate H.M.S. *La Lutine* embarked for Germany with five tons of gold bullion, 14 tons of silver bullion, and three tons of gold and silver coins. In a storm she ran aground on a sandbar between the islands of Terschelling and Vlieland.

18 Scotland Revenues, rents, and taxes collected by Oliver Cromwell from rebellious Scots went down in 1651 when 60 ships were lost during a storm inside muddy, shallow Dundee harbor.

19 Indian Ocean Bound from Lisbon to Goa, India, the Portuguese ship *Conceição* struck a reef near the Peros Banhos Islands in the Chagos Archipelago in 1555. She carried 500 passengers and tons of gold and silver coins. The captain and a few others commandeered the longboat and reached safety, leaving the rest to starve.

20 Philippines The Spanish galleon *San Jeronimo,* bound for Manila with 1½ million pesos in gold and silver, ran onto a reef during a storm off Catanduanes Island near Luzon in 1600.

JONATHAN BLAIR

In shallow water off Motya, Sicily, archeologists salvage the remains of the first Phoenician vessel ever found. "This wreck has provided important clues to the way ships were built in pre-Christian times," says expedition leader Honor Frost. Here she recovers powdery fragments of lead sheathing used by the ancients to protect the hull from wood-boring worms. Carbon-14 dating of timbers from the vessel—apparently a warship—indicates it sank around 250 B.C. An 18-inch-long oak log and a length of rope (opposite) show remarkably little damage after more than 2,000 years in the sea. A diver (opposite, lower) retrieves planking from the vessel's port side. Found in the top layer of sand, the encrusted spade (left) dates from more recent times.

With extreme care, Honor
Frost (foreground) and
Peter Ball remove a fragile
pinewood plank from the
Phoenician ship. Around
her waist Miss Frost wears a
plastic buoy; when inflated
it will help carry heavy
artifacts to the surface. Using
techniques of recovery,
preservation, and recon-
struction developed over
the last three decades,
archeologists will restore
this 2,200-year-old vessel.
The unique find raises
hopes of locating addi-
tional evidence of the
master mariners' far-
ranging voyages. Did the
Phoenicians sail around
the tip of Africa? Did they
reach American shores?
What other priceless relics
of man's past lie buried in
the depths? Future divers
—both archeologists and
treasure hunters—will
provide some of the
answers.

JONATHAN BLAIR

195

Contributors

SUE SWEENEY ABBOTT began combining her interest in sea shells with scuba diving in 1970. Since then she has made frequent diving trips to the seas and oceans of the world, observing and collecting mollusks. In addition, she has worked as a diver on archeological sites in the waters of Greece. Mrs. Abbott, a member of the American Malacological Union, lives in Wilmington, Delaware, with her teen-age sons, Wilson and Barnes Darwin.

Diver-legislator-author ROBERT E. CAHILL is a member of a Massachusetts Task Force on Coastal Resources, and is also co-sponsor of legislation protecting treasure and artifacts found in his state's waters. He dived extensively in the Red Sea while serving with the U. S. Army in East Africa, and later taught scuba diving in Rhode Island. Cahill has written more than 200 articles on undersea adventure, treasure hunting, ecology, and New England history.

NATIONAL GEOGRAPHIC Assistant Editor WILLIAM GRAVES has dived for stories in the waters of Hawaii, Florida, the Bahamas, Bermuda, New England, the Mediterranean, Scandinavia, and Africa. A graduate of Harvard University, he served as a Foreign Service Officer in Germany and Japan, and reported political events for a Washington news agency before joining the GEOGRAPHIC staff in 1956. He is the author of the Special Publication *Hawaii.*

MENDEL PETERSON, a native of Idaho, studied at Vanderbilt University and the Naval Academy Graduate School. He served in the U. S. Navy before joining the staff of the Smithsonian Institution in 1948; from 1969 until his retirement in 1973 he was Director of Underwater Exploration and Curator of Historical Archeology. A collector of ancient and medieval coins, he now acts as consultant in matters relating to underwater exploration and historical materials.

Belgian diver ROBERT STÉNUIT began his career exploring the subterranean rivers and flooded caves of the Ardennes region of Belgium. He was chief diver for Edwin A. Link's Project Man-in-Sea in 1962, once spending 48 hours at a depth of 422 feet in the Bahamas—then the deepest, longest dive ever attempted. Since 1967 he has devoted his time to researching, excavating, studying—and writing about—wreckage of post-medieval ships.

While making a documentary film on Turkish sponge divers in the early 1960's, PETER THROCKMORTON learned of ancient shipwrecks off that coast—and has been involved in underwater archeology ever since. He sailed as a merchant seaman in World War II, and studied anthropology at the Universities of Hawaii, Mexico City, and Paris. He is the author of three books and numerous articles on underwater archeology.

N.G.S. PHOTOGRAPHER ROBERT S. OAKES

Encrusted perfume pitcher, 1715 Spanish fleet.

Index

Boldface indicates illustrations;
italic refers to picture captions.

Library of Congress CIP Data
National Geographic Society, Washington, D. C.   Special Publications Division.
    Undersea treasures.
    1. Treasure-trove.   2. Underwater archeology.
    I. Title.
    G525.N33   1974   910'.453   74-1563

ISBN 0-87044-147-7

Additional References

The reader may want to check the *National Geographic Index* for related articles, and to refer to the Special Publication *World Beneath the Sea* and to the following other books:
    R. Tucker Abbott, *Kingdom of the Seashell*; George F. Bass, editor, *A History of Seafaring: Based on Underwater Archeology and Archaeology Under Water*; Robert F. Burgess, *Sinkings, Salvages, and Shipwrecks*; P. E. Cleator, *Treasure for the Taking*; Bailey W. Diffie, *Latin-American Civilization, Colonial Period*; Nicholas C. Flemming, *Cities in the Sea*; Honor Frost, *Under the Mediterranean*; Sydney J. Hickson, *An Introduction to the Study of Recent Corals*; Dave Horner, *Shipwrecks, Skin Divers and Sunken Gold*; Cyrus H. Karraker, *The Hispaniola Treasure*; Pierre de Latil and Jean Rivoire, *Sunken Treasure*; Robert Marx, *Sea Fever* and *Shipwrecks of the Western Hemisphere*; Alexander McKee, *History Under the Sea*; Robert I. Nesmith, *The Coinage of the First Mint of the Americas at Mexico City, 1536-1572* and *Dig for Pirate Treasure*; J. H. Parry, *The Spanish Seaborne Empire*; Mendel Peterson, *History Under the Sea*; John S. Potter, Jr., *The Treasure Diver's Guide*, second edition; Robert Silverberg, *Sunken History*; Robert Sténuit, *Treasures of the Armada*; Nora Stirling, *Treasure Under the Sea*; Peter Throckmorton, *The Lost Ships*; Frederick Wagner, *Famous Underwater Adventures*; Kip Wagner as told to L. B. Taylor, *Pieces of Eight*.

Acknowledgments

The Special Publications Division is grateful to the individuals and organizations named or quoted in the text and to those cited here for their generous cooperation and assistance during the preparation of this book: Wilburn Cockrell, Florida State Underwater Archeologist; marine biologist Richard W. Grigg, Jack McKenney of *Skin Diver* magazine, and Joseph Rosewater, Division of Mollusks, Smithsonian Institution.

Composition for *Undersea Treasures* by National Geographic's Phototypographic Division, Carl M. Shrader, Chief; Lawrence F. Ludwig, Assistant Chief. Printed and bound by Fawcett Printing Corp., Rockville, Md. Color separation by Chanticleer Co., Inc., New York, N.Y.; Colorgraphics, Inc., Beltsville, Md.; Graphic Color Plate, Inc., Stamford, Conn.; Progressive Color Corp., Rockville, Md.; and J. Wm. Reed Co., Alexandria, Va.